Nanomaterials and Nanocomposites

Engineering Materials

Series Editor
Kaushik Kumar

Nanomaterials and Nanocomposites: Characterization, Processing, and Applications
B. Sridhar Babu and Kaushik Kumar

Sputtered Thin Films: Theory and Fractal Descriptions
Fredrick M Mwema, Esther T Akinlabi, and Oluseyi P Oladijo

ABOUT THE SERIES

This book series focuses on topics of current interest in all aspects of engineering materials. The series aims to be a collection of textbooks, research books, and edited books covering both artificial and natural materials used in engineering and technological applications. It features titles on traditional materials, such as ferrous, nonferrous, ceramic, glass, elastomers, and polymers, as well as advanced materials, including composite materials, smart materials, hierarchical materials, biomaterials, biodegradable materials, functionally graded materials, and shape recognition materials. Topics include macro-, micro-, or nanoscale materials ranging from manufacturing to disposal along with characterization, optimization, applications, and many more.

Experts from industry, research institutes, and academia are invited to submit proposals for books, both authored and edited, on all aspects of materials including but not limited to the above.

For inquires or to submit a book proposal, please contact: Kaushik Kumar, Series Editor (kkumar@bitmesra.ac.in)

Nanomaterials and Nanocomposites
Characterization, Processing, and Applications

Edited by
B. Sridhar Babu and Kaushik Kumar

CRC Press is an imprint of the
Taylor & Francis Group, an **informa** business

First edition published 2021
by CRC Press
6000 Broken Sound Parkway NW, Suite 300, Boca Raton, FL 33487-2742

and by CRC Press
2 Park Square, Milton Park, Abingdon, Oxon, OX14 4RN

© 2021 Taylor & Francis Group, LLC

CRC Press is an imprint of Taylor & Francis Group, LLC

The right of B. Sridhar Babu and Kaushik Kumar to be identified as the author of the editorial material, and of the authors for their individual chapters, has been asserted in accordance with sections 77 and 78 of the Copyright, Designs and Patents Act 1988.

Reasonable efforts have been made to publish reliable data and information, but the author and publisher cannot assume responsibility for the validity of all materials or the consequences of their use. The authors and publishers have attempted to trace the copyright holders of all materials reproduced in this publication and apologize to copyright holders if permission to publish in this form has not been obtained. If any copyright material has not been acknowledged, please write and let us know so we may rectify in any future reprint.

Except as permitted under U.S. Copyright Law, no part of this book may be reprinted, reproduced, transmitted, or utilized in any form by any electronic, mechanical, or other means, now known or hereafter invented, including photocopying, microfilming, and recording, or in any information storage or retrieval system, without written permission from the publishers.

For permission to photocopy or use material electronically from this work, access www.copyright.com or contact the Copyright Clearance Center, Inc. (CCC), 222 Rosewood Drive, Danvers, MA 01923, 978-750-8400. For works that are not available on CCC, please contact mpkbookspermissions@tandf.co.uk

Trademark notice: Product or corporate names may be trademarks or registered trademarks and are used only for identification and explanation without intent to infringe.

Library of Congress Cataloging-in-Publication Data
Names: Babu, B. Sridhar, editor. | Kumar, K. (Kaushik), 1968– editor.
Title: Nanomaterials and nanocomposites : characterization, processing, and applications / edited by B. Sridhar Babu and Kaushik Kumar.
Description: First edition. | Boca Raton, FL : CRC Press, 2021. |
Series: Engineering materials | Includes bibliographical references and index. |
Summary: "This book discusses the most recent research in nanomaterials and nanocomposites for a range of applications, as well as modern characterization tools and techniques. It deals with nanocomposites that contain a dispersion of nanosized particulates and carbon nanotubes in their matrices (polymer, metal, and ceramic). This book enables an efficient comparison of properties and capabilities for these advanced materials, making it relevant for academic research and industrial R&D into nanomaterials processing and application"— Provided by publisher.
Identifiers: LCCN 2020049154 | ISBN 9780367483890 (hardback) |
ISBN 9781003160946 (ebook)
Subjects: LCSH: Nanostructured materials. | Nanocomposites (Materials)
Classification: LCC TA418.9.N35 N25447 2021 | DDC 620.1/15—dc23
LC record available at https://lccn.loc.gov/2020049154

ISBN: 978-0-367-48389-0 (hbk)
ISBN: 978-0-367-75092-3 (pbk)
ISBN: 978-1-003-16094-6 (ebk)

Typeset in Times
by codeMantra

Contents

Preface vii
Editors xi
Contributors xiii

SECTION I State of Art

Chapter 1 Recent Developments in Nanomaterial Applications 3

S. Saravanan, E. Kayalvizhi Nangai, S. V. Ajantha, and S. Sankar

Chapter 2 Some Impact of Nanomaterials in Aerospace Engineering 17

V. Dhinakaran, M. Swapna Sai, and M. Varsha Shree

Chapter 3 Lightweight Polymer–Nanoparticle-Based Composites: An Overview 31

Harrison Shagwira, F.M. Mwema, and Thomas O. Mbuya

Chapter 4 The Role and Applications of Nanomaterials in the Automotive Industry 51

V. Dhinakaran and M. Varsha Shree

Chapter 5 Characterization Tools and Techniques for Nanomaterials and Nanocomposites 61

Ruma Arora Soni, R. S. Rana, and S. S. Godara

SECTION II Synthesis

Chapter 6 Synthesis and Fabrication of Graphene/Ag-Infused Polymer Nanocomposite 87

Chinedu Okoro, Zaheeruddin Mohammed, Lin Zhang, Shaik Jeelani, Zhongyang Cheng, and Vijaya Rangari

SECTION III Analysis

Chapter 7 Aging and Corrosion Behavior of Ni- and Cr-Electroplated Coatings on Exhaust Manifold Cast Iron for Automotive Applications 111

T. Ramkumar, C. A. K. Arumugam, and M. Selvakumar

Chapter 8 Experimental Evaluation of Wear and Coefficient of Frictional Performance of Zirconium Oxide Nanoparticle–Reinforced Polymer Composites for Gear Applications 121

S. Sathees Kumar and B. Sridhar Babu

Chapter 9 Comparative Numerical Analyses of Different Carbon Nanotubes Added with Carbon Fiber–Reinforced Polymer Composite 139

R. Vijayanandh, G. Raj Kumar, P. Jagadeeshwaran, Vijayakumar Mathaiyan, M. Ramesh, and Dong Won Jung

Index 167

Preface

The editors are pleased to present the book *Nanomaterials and Nanocomposites: Characterization, Processing, and Applications* as a part of the *Engineering Materials Series*. This book title was chosen understanding the current importance of *Nanomaterials* as well as familiarization with one of the most sort-out group of materials *Nanocomposites* for industrial and manufacturing world.

The end of the 20th century witnesses a novel evolution of a technology called *Nanotechnology*. It is a technology that deals with very small-sized objects and systems by scheming the structures with the help of nanoscale and also deals with drawing, categorization, construction, appliance of structures, devices at nanoscale etc. The term "nano" means dwarf, is one billionth of a meter. In 1965, the American Physicist Nobel Richard Feynman said *"The Principles of Physics as far as. I can see, do not speak against the possibility of maneuvering things atom by atom."* Feynman's definition was expanded by Drexler *"nanotechnology is the principle of atom manipulation atom by atom through control of the structure of matter at the molecular level. It entails the ability to build molecular systems with atom by atom precision, yielding a variety of nano machines."* A remarkable part of nanotechnology is the incomprehensibly expanded proportion of surface territory to volume present in numerous nanoscale materials, which open up for additional opportunities in surface-based science, for example, catalysis. Nanotechnology has incredible accomplishments and tackles extraordinary issues; however, it will like shrewd present open doors for huge maltreatment. So nanomaterials are generally classified into four categories based on dimensional aspects such as zero-, one-, two-, and three-dimensional particles. Nanomaterials lead to current development in the area of strong technical research, owing to an extensive range of prospective applications such as electronic, optical, and biomedical fields.

Nanocomposites, on the other hand, are composites that contain one of the phases in nanosize (10^{-9} m). These composite materials started to be produced because of their superior physical, thermal, and mechanical properties in comparison with traditional and microcomposites. Besides, the preparation techniques and processing of these nanocomposites show different challenges as a result of the stoichiometry in the nano-phased and elementary structure. Nano-phased filler materials are integrated into the matrix of the composite to enhance the properties of the nanocomposites.

The main aim of this book is to provide an insight about the most recent research in nanomaterials and nanocomposites for a range of applications, modern characterization tools, and techniques. Further, this book deals with the synthesis of nanocomposites with nano-sized particulates in the matrices (polymer, metal, and ceramic). This book also discusses the analysis of the prepared nanocomposites. The primary aim of this book is to open the horizon of the subject to university students studying material science and practicing engineers and professionals on this new group of materials.

The entire book is divided into **3 sections** containing 9 chapters. The **three sections** are as follows: **Section I – State of Art; Section II – Synthesis;** and

Section III – Analysis. Section I contains **Chapters 1–5; Section II** comprises **Chapter 6**; and **Section III** comprises **Chapters 7–9**.

Section I starts with **Chapter 1**, which provides the readers an insight into latest advancement in nanomaterials that can be used in different ways. In this present scenario, the innovations in science and nanotechnologies have made our life much easier and comfortable. The technology involved in nanoscience represents a vital domain in research that involves mechanisms with new properties. Nanomedicine is emergent in the personalized treatments that help in the genome of the patients who are expected to be beneficial out of this. Nanomaterials are crucial in the enhancement of function in addition to the compatibility of implantable medical devices, diagnostic tools, and the system by which drug is delivered. Applications of these nanomaterials towards tissue engineering, implantable device, diagnostic tools, and drug delivery system have been presented in this chapter in detail.

Chapter 2 deals with the impact of nanomaterials and nanotechnology in various sectors of aerospace engineering. The subject requires effective and elaborative discussion in order to obtain enhanced rigidity, stability, strength, cost-efficient products as comparable to conventional metals and other composite materials. Nanomaterials are the elements that possess grain sizes in a nanoscale form in the range of billionth of a meter, and these materials are widely utilized in various applications owing to their exceptional attractive and advantageous properties which can be oppressed in the implication of structural and nonstructural platforms. This chapter includes a specific impact of nanomaterials and nanocomposites in fields of aerospace as the recent technology and its improvement necessitate resources possessing enhanced thermal system with materials consuming great thermal conductivity, mechanical and structural system with resources consuming elevated strength-to-weight ratio properties. The assortment of materials for a specific application plays a major part, and it is very perilous as it desires highly skilled experience with great knowledge due to the variation in properties when operated materials' selection without appropriate acquaintance.

Chapter 3 deals with a specific review of polymer-nanoparticle-based composites with emphasis on the nano-silica reinforcements. The increasing demand for eco-friendly materials in various fields, including the construction industry, has led to increased efforts towards the development of more materials to suit such fields. This chapter provides a background on applications, processing methods, and state of the art. It is noted that there is limited literature focusing on the recycling of polymer using silica nanoparticle-based reinforcements for the construction industry, and hence, gaps in the literature have been identified and the direction for future research focus is presented.

Again, in **Chapter 4**, the effective use of nanomaterials in the automotive industry has been extensively described. A demand for new low-cost, high-performance lightweight materials to replace metals is created by the present global demands for fuel economy and lower emissions from manufacturing and transportation. Nanocomposites are the newly invented class of polymeric materials with superior mechanical, thermal, and processing characteristics that can substitute metals for automotive applications and for any other purpose. In this chapter, the same has been elaborately discussed, to achieve notable rigidity, strength, and reliability comparable or better than metals in the nanocomposite-based components. These materials also possess resistance to degradation, noise damping, improved modulus, thermal stability, and impact resistance.

Chapter 5, the next chapter of this book and the last chapter of Section I, provides the reader with an exclusive study of characterization of engineered nanomaterials. In order to determine the possible risks resulting from their widespread use, characterization of engineered nanomaterials is of vital importance. In fact, nanomaterials have a broad range of physical-chemical properties that have a significant impact on their biological systems interaction. Nanocomposites are new technologies to be studied and used in many applications. The volume fractions of the matrix, the fiber as well as the size and shape of the nanomaterial in the composite can be adapted to nanocomposites. It remained a challenge to prepare nanocomposites with the desired type and scale. The X-ray diffraction (XRD), UV spectroscopy, scanning electron microscope (SEM), and Fourier transform infrared spectroscopy (FTIR) are the characteristic nanocomposites. Thermal analysis (TA) is a valuable way to study a wide variety of polymer properties and can be used for PN to gain more insight into their structure. This chapter highlights various capabilities of TA methods in the emerging field of nanomaterials sciences using the following techniques for nanocomposite material characterization (TMT) applications: thermal gravimetric analysis (TGA), differential calorimetry scanning (DSC), and thermal-mechanical analysis (TMA).

In **Chapter 6**, the only chapter of **Section II**, graphene/silver hybrid (Sg-Ag) nanoparticles were synthesized using a combination of autogenic pressure reactor and microwave irradiation. These graphene/silver hybrid nanoparticles were further infused in epoxy polymer using an ultrasound irradiation to fabricate a conducting polymer nanocomposite. XRD analyses confirmed the crystalline nature of graphene/silver hybrid nanoparticles. Plasticizer (EP9009)-modified epoxy SC-15 resins were loaded with various wt. %age of Sg-Ag nanoparticles. Properties like dielectric constant, thermal effusivity, thermal conductivity, storage modulus, flexure properties, strength, and modulus were found to drastically increase when compared to neat epoxy.

In **Chapter 7**, the next chapter of this book, aimed to investigate the aging and corrosion behavior of exhaust manifold cast iron coated with nickel (Ni) and chromium (Cr) through electrochemical deposition process. The coated manifolds are characterized using SEM. The aging behavior was determined with different stress levels at elevated temperature for different thermal cycles. The corrosion behavior of the coated manifolds was evaluated using weight reduction method and Tafel exploration. Further, the microstructure of the corroded samples was investigated using SEM. A reduction in crack propagation and corrosion was observed and concluded in this chapter.

Chapter 8 reports the formulation, categorization, and experimental determination of zirconium oxide (ZrO2) nanoparticle-reinforced polyamide 6 (PA6) composites. Various test samples were prepared in an injection molding machine, by changing the weight percentage of ZrO2 particles mixed with PA6. Tribological tests were performed on them for different loads, speeds, and ambient conditions on a pin-on disc tribometer. SEM study was used for the study of particle distribution.

The composition was used for the fabrication of gears, which can be used in textile mills, automotive industry, and engineering equipment.

The last chapter of **section III** and this book, i.e., **Chapter 9**, presents comparative numerical analyses of different carbon nanotubes added with carbon fiber-reinforced polymer composite. Generally, epoxy resin plays a vital role in the matrix's selection; therefore, finalization of mixture fundamentally depends on epoxy resin properties. This chapter deals with structural performance of the carbon fiber-based epoxy resin matrix added with various fillers such as carbon nanotubes (single-walled carbon nanotubes and multiwalled carbon nanotubes) by using advanced numerical simulation. ASTM D-3039 provided the geometrical data to ANSYS Design Modeller 16.2 for this successful completion of conceptual design of a test specimen. The discretization and the pure composite generation processes were completed with the help of ANSYS Mesh Tool 16.2 and ANSYS ACP 16.2, respectively. Finally, the comparative analyses in the perspective of various filler additions were executed, and then, the suitable filler is optimized using ANSYS Static Structural. The experimental test and standard theoretical formula were also involved in the structural outputs of the nanocomposite, and the results were validated with numerical simulations.

First and foremost, we would like to thank God. It was your blessing that provided us the strength to believe in passion and hard work, and pursue dreams. We thank our families for having the patience with us for taking yet another challenge, which decreases the amount of time we could spend with them. They were our inspiration and motivation. We would like to thank our parents and grandparents for allowing us to follow our ambitions. We would like to thank all the contributing authors as they are the pillars of this structure. We would also like to thank them to have belief in us. We would like to thank all of our colleagues and friends in different parts of the world for sharing ideas in shaping our thoughts. Our efforts will come to a level of satisfaction if the students, researchers, and professionals concerned with all the fields related to nanomaterials and nanocomposites, in particular, and material science and product development, in general, get benefitted.

We owe a huge thanks to each and every contributing authors, reviewers, editorial advisory board members, book development editor, and the team of **CRC Press** for their availability for work on this huge project. All of their efforts were instrumental in compiling this book, and without their constant and consistent guidance, support, and cooperation, we couldn't have reached this milestone.

Last, but definitely not least, we would like to thank all individuals who had taken time out and help us during the process of writing this book; without their support and encouragement, we would have probably given up the project.

B. Sridhar Babu
Kaushik Kumar

Editors

Kaushik Kumar, B.Tech (Mechanical Engineering, REC (Now NIT), Warangal), MBA (Marketing, IGNOU) and Ph.D (Engineering, Jadavpur University), is presently an associate professor in the Department of Mechanical Engineering, Birla Institute of Technology, Mesra, Ranchi, India. He has 19 years of teaching and research experience and over 11 years of industrial experience in a manufacturing unit of global reputation. His areas of teaching and research interests are composites, optimization, nonconventional machining, CAD/CAM, rapid prototyping, and quality management systems. He has 9 patents, 35+ books, 30 edited books, 55 book chapters, 150 international journal publications, and 22 international and 1 national conference publication to his credit. He is on the editorial board and review panel of 7 international and 1 national journal of reputation. He has been felicitated with many awards and honors. (Web of Science core collection (102 publications/h-index 10+, SCOPUS/h-index 10+, Google Scholar/h-index 23+).

B. Sridhar Babu, professor and dean (IIIC), has completed B.E. (Mechanical Engineering) from Kakatiya University, M.Tech (Advanced Manufacturing Systems) from JNTUH University, and Ph.D (Mechanical Engineering) from JNTUH University. He has 22 years of teaching experience out of which 10 years of experience in CMR Institute of Technology itself. He is fellow of the Institution of Engineers (I), Kolkata, and also member of ISTE, IAENG, and SAE India. He has published 55 papers in various international/national journals and international/national conferences. He is the author of 06 text books. He received the Bharath Jyothi Award for his research excellence from India International Friendship Society, New Delhi, India. He is the reviewer for various international journals and conferences and has guided more than 75 B.Tech and M.Tech projects. His research interests include manufacturing, advanced material, and mechanics of materials. He is a guest editor for Proceedings of 1st International Conference on Manufacturing, Material Science and Engineering (ICMMSE 2019), Materials Today—Proceedings (Scopus and CPCI Indexed), and AIP Proceedings (Scopus Indexed). He is also the guest editor for SN Applied Sciences Springer journal and for 8 edited books.

Contributors

S. V. Ajantha
Department of Physiology
Madha Medical College Hospital and Research Institute
Chennai, India

C. A. K. Arumugam
Department of Mechanical Engineering
Mepco Schlenk Engineering College
Sivakasi, India

B. Sridhar Babu
Department of Mechanical Engineering
CMR Institute of Technology
Hyderabad, India

Zhongyang Cheng
Department of Material Science and Engineering
Auburn University
Auburn, AL

V. Dhinakaran
Department of Mechanical Engineering
Chennai Institute of Technology
Kundrathur, India

S. S. Godara
Department of Mechanical Engineering
RTU
Kota, India

P. Jagadeeshwaran
Department of Mechanical Engineering
Jeju National University
Jeju City, South Korea

Shaik Jeelani
Department of Material Science and Engineering
Tuskegee University
Tuskegee, AL

Dong Won Jung
Department of Mechanical Engineering
Jeju National University
Jeju City, South Korea

G. Raj Kumar
Rajalakshmi Institute of Technology
Chennai, India

S. Sathees Kumar
Department of Mechanical Engineering
CMR Institute of Technology
Hyderabad, India

Vijayakumar Mathaiyan
Department of Aeronautical Engineering
Kumaraguru College of Technology
Coimbatore, India

Thomas O. Mbuya
Department of Mechanical & Manufacturing Engineering
University of Nairobi
North, Kenya

Zaheeruddin Mohammed
Department of Material Science and Engineering
Tuskegee University
Tuskegee, AL

F. M. Mwema
Department of Mechanical Engineering
Dedan Kimathi University of Technology
Nyeri, Kenya

E. Kayalvizhi Nangai
Department of Physics
K. Ramakrishnan College of Technology
Trichy, India

Chinedu Okoro
Department of Material Science and Engineering
Tuskegee University
Tuskegee, AL

M. Ramesh
UG Student
Department of Aeronautical Engineering
Kumaraguru College of Technology
Coimbatore, Tamil Nadu, India

T. Ramkumar
Department of Mechanical Engineering
Dr. Mahalingam College of Engineering and Technology
Pollachi, India

R. S. Rana
Department of Mechanical Engineering
Maulana Azad National Institute of Technology
Bhopal, India

Vijaya Rangari
Department of Material Science and Engineering
Tuskegee University
Tuskegee, AL

M. Swapna Sai
Centre for Applied Research
Chennai Institute of Technology
Kundrathur, Chennai

S. Sankar
Department of Semiconductor Science
Dongguk University – Seoul
Seoul, Republic of Korea.

S. Saravanan
Department of Mechanical Engineering
K. Ramakrishnan College of Technology
Trichy, India

M. Selvakumar
Department of Automobile Engineering
Dr. Mahalingam College of Engineering and Technology
Pollachi, India

Harrison Shagwira
Department of Mechanical Engineering
Dedan Kimathi University of Technology
Nyeri, Kenya

M. Varsha Shree
Centre for Applied Research
Chennai Institute of Technology
Kundrathur, Chennai

Ruma Arora Soni
Energy Centre
Maulana Azad National Institute of Technology
Bhopal, India

R. Vijayanandh
Department of Aeronautical Engineering
Kumaraguru College of Technology
Coimbatore, India

Lin Zhang
Department of Material Science and Engineering
Auburn University
Auburn, AL

Section I

State of Art

1 Recent Developments in Nanomaterial Applications

S. Saravanan and E. Kayalvizhi Nangai
K. Ramakrishnan College of Technology

S. V. Ajantha
Madha Medical College Hospital and Research Institute

S. Sankar
Dongguk University Seoul V. Dhinakaran
Chennai Institute of Technology

CONTENTS

1.1 Introduction 3
1.2 Nanomedicine 4
1.3 Application of Nanomaterials in Medicine 6
    1.3.1 Nanomaterials in Tissue Repair and Regeneration 6
    1.3.2 Nanomaterials in Implantation 7
    1.3.3 Nanomaterials in Medical Devices...8
    1.3.4 Nanomaterials in Diagnostic Tools...9
    1.3.5 Nanomaterials in Pharmaceutics 10
1.4 Conclusion 12
References..12

1.1 INTRODUCTION

At the end of the 20th century, a wonderful new technology, called nanotechnology, has emerged. It is a technology that deals with very small-sized objects and systems by scheming the structures at the nanoscale and also with drawing, categorization, construction, and application of structures and devices at the nanoscale [1–3]. The term "nano" means dwarf, and a nanometer is one-billionth of a meter. In 1965, American physicist Nobel Richard Feynman distinguished some useful concepts in nanotechnology and stated that "the principles of physics, as far as I can see, do not speak against the possibility of maneuvering things atom by atom" [4]. Feynman's definition was expanded by Drexler's quote: "nanotechnology is the principle of atom manipulation atom by atom, through control of the structure of matter at the molecular level. It entails the ability to build molecular systems with atom by atom precision, yielding a variety of nanomachines [5]. Binning and Rohrer expounded on Drexler's hypotheses in a handy manner. In 1981, they were the first to observe the particles to investigate nanotechnology. Researchers had the opportunity to get a handle on and prepare the particles for building structures. A remarkable feature of nanotechnology is the incomprehensibly expanded proportion of surface area to volume present in numerous nanoscale materials, which opens up additional opportunities in surface-based science, for example, catalysis. Nanotechnology has incredible accomplishments and tackles extraordinary issues; however, it will be like shrewd open doors for huge maltreatment. In nanomaterials, a maximum number of atoms are situated on the surface of the elements, so it has almost all the increasing surface area. They are generally classified into four categories based on the dimensional aspect such as zero-, one-, two-, and three-dimensional particles. The various dimensions of nanoparticles are shown in Figure 1.1 [6]. Nanomaterials lead to current development in the area of strong technical research, owing to an extensive range of prospective applications such as electronic, optical, and biomedical applications.

Most biological molecules and structures are of a similar size to that of nanomaterials. Consequently, nanomaterials can be used for research and applications in both in vivo and in vitro biomedical fields. Hence, the combination of nanomaterials and biology plays a key role in the enhancement of diagnostic devices along with tools and drug delivery system [7–10]. Biological tests are becoming faster, more aware, and flexible to measure the occurrence or action of chosen substances, while at the nanoscale, particles are set to work as tags or labels. To label these detailed molecules, structures using magnetic nanoparticles are bound to a suitable antibody. Genetic sequence in a model is detected by gold nanoparticles tagged by short segments of DNA [11]. Nanopore technology is used to analyze nucleic acids altering strings of nucleotides openly into electronic signatures. Costs and human suffering can be reduced by this highly selective approach. Dendrimers and nanoporous materials can hold tiny drug molecules transporting them to the preferred place. One more insight is based on small electromechanical systems. Nano electromechanical systems (NEMS) are being examined for the dynamic release of drugs. Iron nanoparticles or gold shells are used for important applications including cancer treatment [12–15]. Recent pharmacological molecular entities are discovered by the selection of pharmaceuticals for specific people to maximize the effectiveness and minimize side effects, and drugs are delivered at targeted locations or tissues inside the body. Nanoparticles can render targeted and constant delivery of biological components to particular tissues with the least systematic side effects [16, 17].

1.2 NANOMEDICINE

The aspect of approaching health care in which explicit medicines for every patient are created in consideration of ecological, phenotypic, and genetic factors is known as personalized medicine. These factors have significantly affected the adequacy and security of the treatment. Nanomaterials have been utilized in the field of medicine for more than 130 years; particularly, colloidal silver was utilized for the anticipation of eye diseases, which in present used in medicine as one of the first nanomedicines and iron dextran. Nanomedicine is the use of nanoscale materials such as engineered

Developments in Nanomaterial Applications

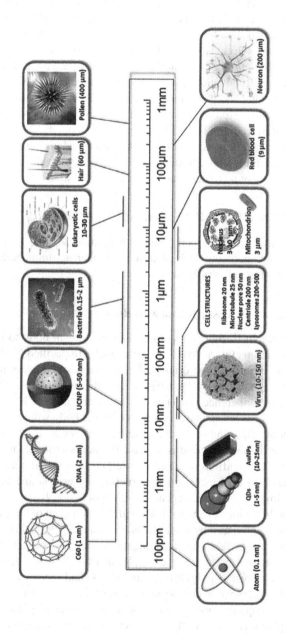

FIGURE 1.1 Size indications of nanomaterials.

nanodevices and nanostructures for preventing, screening, renovating, creating, and running human biological systems [18]. It focuses on the key enabling technologies such as molecular nanotechnology and molecular manufacturing. Human body requires effective catch-up of medicine. The result is the ability to examine and modify the human body entirely like fixing a machine these days [19]. New industrial revolution will be created based on the nano-concepts, but scientists and engineers from various fields work jointly to accomplish the vision. Nanorobots play a significant role in the prevention, diagnosis, and treatment of illnesses [20]. Drug delivery has become a research hot spot in the field of nanomedicine. RNA interference therapy is yet experimental and problematic because of its newness and also the lack of bioavailability. The tiny-sized cells take up lipid or polymer-based nanoparticles instead of being cleared from the body. Drug delivery systems should minimize side effects and possibly reduce both dose and dose frequency and improve the efficiency [21,22]. The pharmaceutical industry functioning further personally is expected to recognize the possibility of nanomedicine for incurable diseases.

1.3 APPLICATION OF NANOMATERIALS IN MEDICINE

This section briefly explains the broad range of nanomaterials that have been used for applications in nanomedicine such as tissue engineering, implantable devices, diagnostic tools, and drug delivery system.

1.3.1 NANOMATERIALS IN TISSUE REPAIR AND REGENERATION

Nanomaterials are designed to be compatible with the human body to replace and repair tissues. Bone and teeth are "hard" tissues which are pacified by reproducing tissues that are indifferent from the original. On a poor tissue implant interface, to overcome this, coating is necessary for different metallic implant materials. Implant design can increase the adherence properties of the natural and implant tissues [23]. "Soft" repair damaged tissues can be self-repaired by the body which results in scar formation on the body, skin, and other tissue can be replaced by graft material [24]. To restore regenerate tissue, a scaffold is necessarily bioresorbable that will act as a temporary structure. The scaffold material is essential for repairing and regenerating damaged tissues [25]. Advances in nanostructure production and improvement have a significant impact on tissue regeneration scaffolds. Polymers are explored based on the optical behavior of nanoparticles with the influence of hybrid scaffolds. Molecular imprints are prepared by using the nanotechnology for maximizing durable feasibility. Nanomaterial fabrication techniques are being investigated essentially for growing large complex organs. For example, heart valves are fabricated from nanopolymer materials such as polyvinyl alcohol (PVA) and seeded with fibroblasts and endothelial cells. Transparent composite hydrogels are made from PVA and subjected to in vitro biocompatibility evaluation with human corneal epithelia cells [26].

Nanomaterials play a vital role in the enhancement of function and compatibility of implantable medical devices. It will contain nanoscale materials with non-intrusive or minimally intrusive systems and smaller nanoscale systems. Functional electrical stimulation is a boon to the physically deformed people who lost their legs.

Developments in Nanomaterial Applications

This method is executed for the physically deformed people to energize the powerless limbs. The muscle fiber membrane is incorporated with potential-generating nanostructures which are increased with membrane permeability and improves the extracellular electrical stimulation. The various potential applications of nanomaterials for bone tissue engineering and bone implantations are shown in Figure 1.2 [27].

1.3.2 Nanomaterials in Implantation

Nano-enabled technologies to provide a mixture of new huge surface area and ability to design more biocompatible nanomaterials and the purpose of coatings on implants are to raise the adhesion, stability, and lifetime. Ceramics such as calcium phosphate are extensively used for implant coatings and are made up of particles of nano-size. The nanomaterial properties can be maintained with the help of new low-temperature processes in electromagnetic fields. Implant nanomaterial coatings are being evaluated for improving the interface, and they can greatly improve the life of humans who need them. Calcium phosphate apatite (CPA) and hydroxyapatite (HAP) nanoceramics have high strength which are most adaptable in consideration with least side effects [28].

Recent developments of the biodegradable materials for bone repairing use provisional biosorbable structures. The necessary porous structure of bone biomaterials contains interrelated pores to admit body fluids among soft and hard tissues. Bones are required for repair or replacement due to bone rupture, splice, dental applications, and other types of surgery. Synthetic bone cement is used for filling bone cavities.

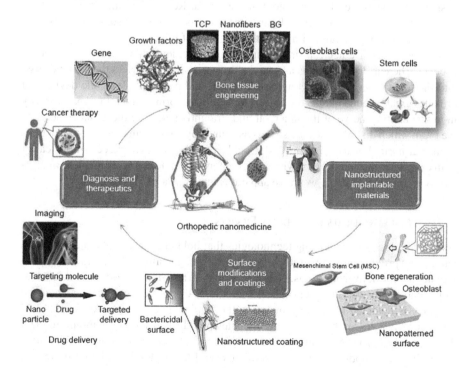

FIGURE 1.2 Nanomaterials for bone tissue and implantation applications.

The bone cement contains PMMA, which acts as a filler. Poly (methyl methacrylate) is injected as flow able glue and then solidifies in vivo [29]. Bioresorbable polymers are biodegradable which offer great potential controlled drug delivery and many medical applications. For example surgical sutures also known as bone addiction and repair devices [30]. A smart material in the body to surpass the present human behavior through integration technologies becomes possible and detects environmental conditions beyond current human limits [31].

Nanotechnologies and its related micro technologies for vision gives fundamentally restructure the technology and to build up of lesser size and most controlling devices to restore missing visualization and hearing task. The devices gather and make over data and convert light into electrical signals which are send to the human nervous system. Two most prevalent retinal degenerative diseases are retinitis pigmentosa and age-related macular degeneration. Retinitis pigmentosa causes progressive failure of photoreceptors and diminishing peripheral vision. This condition often leads to blindness. The neural wiring from the eye to the brain is still intact, and retinal nerves remain intact and functional, but the eyes lack photoreceptor activity, bridging, and, to stimulate adjacent whole cells, could compensate for photoreceptor loss artificially [32].

Retinal implants are devices that are designed to restore vision. They stimulate functional neurons in the retina electrically to restore vision. The artificial retina devices are provided with tiny camera fixed in eyeglasses which confine visual pictures and air sends the message to a microcomputer wear on a belt and transmits it to receiver on the eye. Optobionics create chip which is inserted behind the retina and designed to substitute photoreceptors in the retina. Retinal implant devices initiate electrical simulation and than to light stimulation, so that the visual system is activated and it can significantly improve their quality of life [33].

Patients who have acute hearing impairment have deficiency sensory cells in the cochlea. The ear drum vibrates as sound waves arrive and transfers the sound energy into the middle ear. Cochlear implant is a small and powerful electronic device that provides greater sound quality. Cochlear implants transform sound into signals directly in to the auditory nerve of the inner ear and straightly stimulate the auditory nerve to the brain which distinguishes the signals as sound. By passing any damaged structures in the ear normal hearing is impeded. Cochlear implants maybe placed in one ear or both ears. Tiny microprocessor with a microphone is connected with cochlear implant which is constructed into a wearable apparatus that clips after the ear [34].

1.3.3 Nanomaterials in Medical Devices

Nanotechnology offers sensing technologies that hold enormous potential for health care. It can provide more accurate information for diagnosing disease and more effective details for delivering drugs. Nanotechnology offers new important tools such as implantable and wearable sensing technologies. A specified physical or chemical property is detected effectively by using the implantable sensors. At Texas A&M and Penn State researchers who have diabetes blood sugar levels are monitored [34, 35]. Optical micro sensors are one type of sensors. After surgery, tissue circulation is monitored by optical micro sensors that would be implanted into sub dermal or deep tissue device is connected with data transmission to close by sensor. Paralyzed limbs should be treated and monitored by using these sensors. To determine normal and problematic data, strain, acceleration, angular rate, and related parameters can be measured using implantable MEMS sensors [36–39].

Nanomaterials are being utilized to improve the role of the surgeon. Verimetra is developing an enhanced version of its Data Knife with logic and surgical microelectromechanical systems that can provide added information and functionality to assist a surgeon during a procedure by stimulating electrodes, measuring and cutting with ultrasonic elements, and cauterizing. Instruments are being developed with specific functionality such as tilt and pressure to allow operating tasks to be performed by neurosurgeons with greater precision and safety. Nanoparticles are also being investigated for optically guiding surgery. This can potentially allow for better removal of lesioned or diseased sites, including tumors [40–42].

Robotic surgical systems enable the surgeon to perform minimally invasive surgery with an advanced set of instruments. Robotic systems can be used to enhance the surgeon's ability, precision, flexibility, and control during the operation. Compared with traditional techniques, they may allow the surgeon to better view the surgical site. Minimally invasive surgery is a less stressful procedure and three-dimensional high definitions are present in the view of surgical area. The surgeon who performs operations sits at the robotic console operating in the control system without any aid of surgical instruments. Figure 1.3 illustrates nanorobots' main design features and classification. Harmful bacteria are collected from the mouth by tooth-cleaning nanorobots. A cream that contains nanorobots may be used to cure skin diseases. A right quantity of dead skin and excess oils might possibly be removed using nanorobots [43].

1.3.4 NANOMATERIALS IN DIAGNOSTIC TOOLS

The speed and accuracy of recognizing genes can be increased using nanomedicine, and it gives new solutions of genetic materials in drug discovery and development. Several new technologies improve the target identification of genes. Gold nanoparticle probes are being interacted with chemicals that adhere to targeted genetic materials and enlight the sample which is away to light [44]. Nanodevices provide high-quality images with new methods of treatment, which not possible with the current devices.

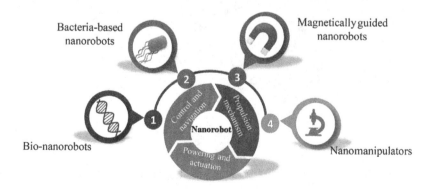

FIGURE 1.3 Main design features and classification of nanorobots.

Malignant tumors are detected and contained at the premature period of their growth. The earlier a tumor is detected, the more likely a successful treatment outcome and surgery can be effectively removing the tumor. Nanotechnology recommends a novel set of tools and solutions in favor of early detection of cancer and other diseases. Miniature wireless devices are being designed at the nano-level. Traditional devices did not produce high-quality images compared with micro technology. Improved imaging with better contrast agents helps to diagnose diseases more sensitively [45].

Cancer cells are attracted by magnetic nanoparticles added to a cancer antibody. The nanoparticles are also combined with a dye, which is extremely noticeable on an MRI. Nanoparticles are used as targeting agents for cancer therapy comprised anticancer drugs as shown in Figure 1.4. The nano-sized particles were carried out and circulated through the bloodstream. Chemotherapy is unconfined to the capillary membrane after attachment. Additional cancer sites were detected by the nanoparticles that are traveled in the bloodstream [46–50].

1.3.5 Nanomaterials in Pharmaceutics

Nanomaterials will have a high impact on the pharmaceutical industries and their development strategy by providing solutions to decrease the discovery and potentially reduce the time for developing new therapies and development costs.

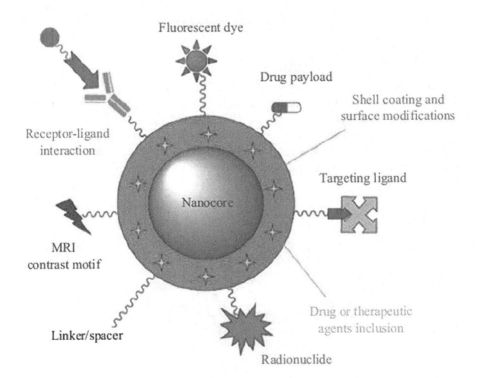

FIGURE 1.4 Nanoparticles used as targeting agents for cancer therapy.

Developments in Nanomaterial Applications

Nanomaterials offer a broad range of new materials that optimize the release of pharmaceutical goods for developing customized solutions.

Figure 1.5 shows the various classifications of nanomedicines for drug delivery system. Various drug delivery approaches can be used to increase the therapeutic performance and reduce side effects. Drug encapsulation materials include liposomes and polymers, which are used as micro scale particles. Tiny particles are surrounded by a coating to give small capsules and these drugs can be released at a particular time through the encapsulated material that degrades in the body. At the nanoscale, materials exhibited greater properties than at the micro scale for certain drug delivery challenges [51]. Encapsulation of drugs in nanoparticles is being investigated for curing neurological disorders. The central nervous system would be treated by delivered therapies across the blood–brain barrier. Neurotech is used as a semi permeable membrane that permits the diffusion by therapeutic agents through the membrane of encapsulated cells. Cells are isolated and antibody rejections are minimized by the membrane. Nanoparticles of biodegradable polymer coated with poly (butyl cyano acrylate) considerably enhance the anti-tumor effect of doxorubicin [52].

Nanomaterial drug delivery system has functional properties that carry drugs to their destination sites. Certain nanostructures can deliver drugs to the targeted sites and attract specific cells and then control release of drugs by crossing the barrier of the living systems when required. Fullerenes, dendrimers, and Nanoshells are smart nanostructures majorly being used in cancer treatment. Fullerenes are the allotropy

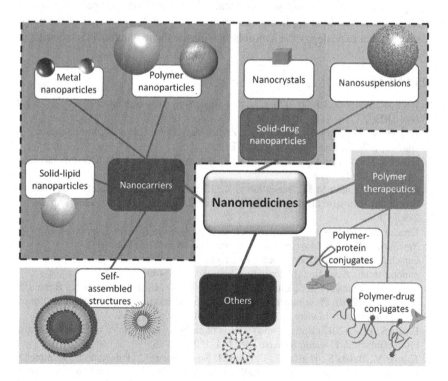

FIGURE 1.5 Different categories of nanomedicines for drug delivery system.

form of carbon, and their molecules are hollow spherical. They contain C_{60} carbon atoms and are referred to as buckminsterfullerene/ buck balls. They can be used for the drug delivery system in the body. They can act as hollow cages to entrap other atoms or molecules, and fullerene derivatives are attached to the targeting agents. Fullerene-based drug delivery platforms are being developed using C_{60} atoms that connect through antibodies [53]. Scientists employ rational drug design methods which have been used in fullerenes platform technology and C_{60} which has fashioned some drug applicants by HIV/AIDS, radical scavenger, and antioxidant [54].

1.4 CONCLUSION

This chapter has contributed well to the review of existing research on nanomaterials in medicine. A systematic framework on the review of nanomedicines with nanomaterial applications is explained in the introductory part. Among the applications of nanomaterials, the importance of nanoparticles in medicine is discussed in detail. Section 1.3 discusses the preliminaries in five parts, namely tissue repair and regeneration, implantation, medical devices, diagnostic tools, and pharmaceutics of nanomaterials. Accordingly, motivated by normal living beings and relying upon the application site and target cell, real nanorobots relied upon a huge effect on the human beings with numerous ailments. Even though nanorobots have been designed by scientific professionals related to the field of medicine also, this field is lagging in manufacturing technology that is not yet well established. This domain needs to draw the attention in the biomedical applications to reach every nook and corner of the world. It is expected that in posterity the nanorobots will rule the medical field in abundance.

REFERENCES

1. Wu L.-P., Wang D., Li Z. (2020) Grand challenges in nanomedicine. *Mater. Sci. Eng. C* 106: 110302.
2. Wagner V., Dullaart A., Bock A.-K., Zweck A. (2006) The emerging nanomedicine landscape. *Nat. Biotechnol.* 24: 1211–1217.
3. National Nanotechnology Initiative (NNI). Available online: www.nano.gov (accessed on 22 July 2019).
4. Hussein A., Zagho M.M., Nasrallah G.K., Elzatahry A.A. (2018) Recent advances in functional nanostructures as cancer photothermal therapy. *Int. J. Nanomedicine* 13: 2897–2906.
5. Drexler E.K., Peterson C., Pergamit G. *Unbounding the Future: The Nanotechnology Revolution.* William Morrow and Company, Inc.: New York, NY, 1991.
6. Gnach A., Lipinski T., Bednarkiewicz A., Rybka J., Capobianco J.A. (2015) Upconverting nanoparticles: Assessing the toxicity. *Chem. Soc. Rev.* 44: 1561–1584.
7. Li Q., Ohulchanskyy T.Y., Liu R., Koynov K., Wu D., Best A., Kumar R., Bonoiu A., Prasad P.N. (2010) Photoluminescent carbon dots as biocompatible nanoprobes for targeting cancer cells in vitro. *J. Phys. Chem. C* 114: 12062–12068.
8. Bayda S., et al. (2017) Bottom-up synthesis of carbon nanoparticles with higher doxorubicin efficacy. *J. Control. Release* 248: 144–152.
9. Kumar V., Bayda S., Hadla M., Caligiuri I. Russo Spena C., Palazzolo S., Kempter S., Corona G., Toffoli G., Rizzolio F. (2016) Enhanced chemotherapeutic behavior of open-caged DNA@doxorubicin nanostructures for cancer cells. *J. Cell. Physiol.* 231: 106–110.

10. Yuan Y., Gu Z., Yao C., Luo D., Yang D. (2019) Nucleic acid–based functional nanomaterials as advanced cancer therapeutics. *Small* 15: 1900172.
11. Palazzolo S., et al. (2019) Proof-of-concept multistage biomimetic liposomal DNA origami nanosystem for the remote loading of doxorubicin. *ACS Med. Chem. Lett.* 10: 517–521.
12. Kumar V., Palazzolo S., Bayda S., Corona G., Toffoli G., Rizzolio F. (2016) DNA nanotechnology for cancer therapy. *Theranostics* 6: 710–725.
13. Sharma N., Sharma M., Sajid Jamal Q.M., Kamal M.A., Akhtar S. (2019) Nanoinformatics and biomolecular nanomodeling: A novel move en route for effective cancer treatment. *Environ. Sci. Pollut. Res. Int.* 20(16): 1–15.
14. Liu J., Chen Q., Feng L., Liu Z. (2018) Nanomedicine for tumor microenvironment modulation and cancer treatment enhancement. *Nano Today* 21: 55–73.
15. Park W., Heo Y.J., Han D.K. (2018) New opportunities for nanoparticles in cancer immunotherapy. *Biomater Res.* 22: 24.
16. Choi Y.H., Han H-K. (2018) Nanomedicines: Current status and future perspectives in aspect of drug delivery and pharmacokinetics. *J. Pharm. Investig.* 48(1): 43–60.
17. Farjadian F., Ghasemi A., Gohari O., Roointan A., Karimi M., Hamblin M.R. (2019) Nanopharmaceuticals and nanomedicines currently on the market: Challenges and opportunities. *Nanomedicine* 14(1): 93–126.
18. Pardi N., Hogan M.J., Porter F.W., Weissman D. (2018) mRNA vaccines—A new era in vaccinology. *Nat. Rev. Drug. Discov.* 17(4): 261–279.
19. Zhang C., Maruggi G., Shan H., Li J. (2019) Advances in mRNA vaccines for infectious diseases. *Front Immunol.* 10: 594.
20. Lundstrom K. (2018) Latest development on RNA-based drugs and vaccines. *Future Sci. OA.* 4(5): FSO300.
21. Kumar A., Patel A., Duvalsaint L., Desai M., Marks E.D. (2014) Thymosin β4 coated nanofiber scaffolds for the repair of damaged cardiac tissue. *J. Nanobiotechnol.* 12: 10.
22. Cohen-Karni T., Lieber C.M. (2013) Nanowire nanoelectronics: Building interfaces with tissue and cells at the natural scale of biology. *Pure Appl. Chem.* 85: 883–901.
23. Gorain B., Tekade M., Kesharwani P., Iyer A.K., Kalia K., Tekade R.K. (2017) The use of nanoscaffolds and dendrimers in tissue engineering. *Drug Discov. Today* 12(4): 652–664.
24. Fleischer S., Shevach M., Feiner R., Dvir T. (2014) Coiled fiber scaffolds embedded with gold nanoparticles improve the performance of engineered cardiac tissues. *Nanoscale* 6: 9410–9414.
25. Kettiger H., Schipanski A., Wick P., Huwyler J. (2013) Engineered nanomaterial uptake and tissue distribution: From cell to organism. *Int. J. Nanomed.* 8: 3255–3269.
26. Wu L.P., Wang D., Parhamifar L., Hall A., Chen G.Q., Moghimi S.M. (2014) Poly (3-hydroxybutyrate- co-R-3-hydroxyhexanoate) nanoparticles with polyethylenimine coat as simple, safe, and versatile vehicles for cell targeting: Population characteristics, cell uptake, and intracellular trafficking. *Adv. Healthc. Mater.* 3: 817–824.
27. Shastri V.P. (2003) Non-degradable biocompatible polymers in medicine: Past, present and future. *Curr. Pharm. Biotechnol.* 4(5): 331–337.
28. Feng S.-S., Mu L., Win K.Y., Huang G. (2004) Nanoparticles of biodegradable polymers for clinical administration of paclitaxel. *Curr. Med. Chem.* 11: 413–424.
29. Duan X., Xiao J., Yin Q., Zhang Z. (2013) Smart pH-sensitive and temporal-controlled polymeric micelles for effective combination therapy of doxorubicin and disulfiram. *ACS Nano* 7(7): 5858–5869.
30. Maisano M., Cappello T., Catanese E., Vitale V., Natalotto A., Giannetto A., Barreca D., Brunelli E., Mauceri A., Fasulo S. (2015) Developmental abnormalities and neurotoxicological effects of CuO NPs on the black sea urchin Arbacia lixula by embryotoxicity assay. *Mar. Environ. Res.* 111: 121–127.

31. Órdenes-Aenishanslins N.A., Saona L.A., Durán-Toro V.M., Monrás J.P., Bravo D.M., Pérez-Donoso J.M. (2014) Use of titanium dioxide nanoparticles biosynthesized by *Bacillus mycoides* in quantum dot sensitized solar cells. *Microb. Cell. Factories* 13: 90.
32. Wennerberg A., Jimbo R., Allard S., Skarnemark G., Andersson M. (2011) *In vivo* stability of hydroxyapatite nanoparticles coated on titanium implant surfaces. *Int. J. Oral. Maxillofac. Implants* 26: 1161–1166.
33. DiSanto R.M., Subramanian V., Gu Z. (2015) Recent advances in nanotechnology for diabetes treatment. *Wiley Interdiscip. Rev. Nanomed. Nanobiotechnol.* 7(4): 548–564.
34. Wu L.-P., Ficker M., Mejlsøe S.L., Hall A., Paolucci V., Christensen J.B., Trohopoulos P.N., Moghimi S.M. (2017) Poly-(amidoamine) dendrimers with a precisely core positioned sulforhodamine B molecule for comparative biological tracing and profiling. *J. Control. Release* 246: 88–97.
35. Cheng I.F., Chang H.C., Chen T.Y., Hu C., Yang F.L. (2013) Rapid (<5 min) identification of pathogen in human blood by electrokinetic concentration and surface-enhanced Raman spectroscopy. *Sci. Rep.* 3: 2365.
36. Washington D.C., Karimi M., Mirshekari H., Basri S.M.M., Bahrami S., Moghoofei M., Hamblin M.R. (2016) Bacteriophages and phage-inspired nanocarriers for targeted delivery of therapeutic cargos. *Adv. Drug Deliv. Rev.* 106: 45–62. https://doi.org/10.1016/j.addr.2016.03.003
37. Hariadi R.F., Sommese R.F., Adhikari A.S., Taylor R.E., Sutton S., Spudich J.A., Sivaramakrishnan S. (2015) Mechanical coordination in motor ensembles revealed using engineered artificial myosin filaments. *Nat. Nanotechnol.* 10: 696–700.
38. Sun R., Wang W., Wen Y., Zhang X. (2015) Recent advance on mesoporous silica nanoparticles-based controlled release system: Intelligent switches open up new horizon. *Nano* 5: 2019–2053.
39. Li J., et al. (2017) Micronanorobots for biomedicine: Delivery, surgery, sensing, and detoxification. *Sci. Robot* 2(4): 194–215
40. Li S., et al. (2018) A DNA nanorobot functions as a cancer therapeutic in response to a molecular trigger in vivo. *Nat. Biotechnol.* 36(3): 258.
41. Wan Z., Wang Y., Li S.S., Duan L., Zhai J. (2005) Development of array-based technology for detection of HAV using Gold-DNA probes. *J. Biochem. Mol. Biol* 38: 399–406.
42. Tsai T.-T., Huang C.-Y, Chen C.-A., Shen S.-W., Wang M.-C., Cheng C.-M., Chen C.-F. (2017) Diagnosis of tuberculosis using colorimetric gold nanoparticles on a paper-based analytical device. *ACS Sens.* 2: 1345–1354.
43. Wu C.H., Kuo Y.H., Hong R.L., Wu H.C. (2015) Alpha-Enolase-binding peptide enhances drug delivery efficiency and therapeutic efficacy against colorectal cancer. *Sci. Transl. Med.* 7(290): 290-91.
44. Ashton S., et al. (2016) Aurora kinase inhibitor nanoparticles target tumors with favorable therapeutic index in vivo. *Sci. Transl. Med.* 8(325): 325-17.
45. Nicolas S., et al. (2018) Polymeric nanocapsules as drug carriers for sustained anticancer activity of calcitriol in breast cancer cells. *Int. J. Pharm.* 550(1–2): 170–179.
46. Freire C., Ramos R., Puertas R., Lopez-Espinosa M.J., Julvez J., Aguilera I., Cruz F., Fernandez M.F., Sunyer J., Olea N. (2010) Association of traffic-related air pollution with cognitive development in children. *J. Epidemiol. Commun. Health.* 64: 223–228.
47. Wicki A., Witzigmann D., Balasubramanian V., Huwyler J. (2015) Nanomedicine in cancer therapy: Challenges, opportunities, and clinical applications. *J. Control. Release* 200: 138–157.
48. Lin Z.C., McGuire A.F., Burridge P.W., Matsa E., Lou H.-Y., Wu J.C., Cui B. (2017) Accurate nanoelectrode recording of human pluripotent stem cell-derived cardiomyocytes for assaying drugs and modelling disease. *Microsyst. Nanoeng.*, 13(3): 16080.
49. Torchilin V.P. *Passive and Active Drug Targeting: Drug Delivery to Tumours as an Example, Drug Deliv.* Springer, USA, 2010, pp. 3–53.

50. Bilan R., Nabiev I., Sukhanova A. (2016) Quantum dot-based nanotools for bioimaging, diagnostics, and drug delivery. *ChemBioChem* 17: 2103–2114.
51. Skotland T., Iversen T.G., Torgersen M.L., Sandvig K. (2015) Cell-penetrating peptides: Possibilities and challenges for drug delivery in vitro and in vivo. *Molecules* 20: 13313–13323.
52. Doane T.L., Burda C. (2012) The unique role of nanoparticles in nanomedicine: Imaging, drug delivery and therapy. *Chem. Soc. Rev.* 41: 2885–2911.
53. Ventola C.L. (2017) Progress in nanomedicine: Approved and investigational nanodrugs. *P. & T: Peer-Reviewed J. Formul. Manag.* 42: 742.
54. Kumar A., Mazinder Boruah B., Liang X.-J. (2011) Gold nanoparticles: Promising nanomaterials for the diagnosis of cancer and HIV/AIDS. *J. Nanomater.* 202187: 1–17.

2 Some Impact of Nanomaterials in Aerospace Engineering

V. Dhinakaran, M. Swapna Sai, and M. Varsha Shree
Chennai Institute of Technology

CONTENTS

2.1 Introduction 17
2.2 Nanomaterial Structure 19
2.3 Surface Properties of Nanomaterials and CNTs in Aerospace 19
2.4 Microscopy Methods of Nanomaterials 20
2.5 Scanning Tunneling Microscope (STM) 21
2.6 Atomic Force Microscopy (AFM) 22
    2.6.1 Radar Absorbing Materials 22
    2.6.2 Nanochassis 22
2.7 Prominence of Nanomaterials in Aerospace Industry 22
2.8 CNT Structures for Aerospace Components 25
    2.8.1 Future Scope 25
    2.8.2 Cons of Nanomaterials 26
    2.8.3 Conclusion 26
References 27

2.1 INTRODUCTION

Nanomaterials are now a large industry because of their superior synthesis processes and increased control, and new properties of materials developed at the nanoscale indicate that this field is evolving with micronutrients as the mixture of reinforced nanoparticles in the source material. Nanoparticles are no more than a dry form of solid metal in small quantities [1]. They are actually very small in size (1 nm = 10^{-9} m) and possess elevated electrical, thermal, corrosion resistance, and mechanical properties. They have a large surface area, and their surface-to-volume ratio property enhances the rate of heat transfer as the available surface area increases. Much research is being done in the field of behavioral analysis of nanoparticles in a variety of situations or applications [2]. The nanomaterials of various forms such as nanoparticles, nanofibers, and nanofilms are extensively employed in various manufacturing sectors such as energy systems, construction of parts, biomedical devices, chemicals, electronic sensors, agricultural fields, aerospace, automotives, paints, and cosmetics. The process of re-engineering of materials and its equipment is well known to be nanotechnology by monitoring substance at the molecular range [3]. Nanotechnology is the strategy, manufacturing technique and involves the application of nanoscale materials in modern macro- and microsystems by considering the basic associations among structural properties and materials. It works on nanometer-scale materials and innovative science from micrometers to several macrometers [4]. The use of nanomaterials in various aircraft components is represented in Figure 2.1. The production of nanomaterials and nano-sized structures is an important aspect of nanotechnology, and its true application is probable only when nanostructured materials with the desired properties such as size, chemical composition, morphology, and physical behavior are available. The manufacture of micronutrients began a long time ago, but nanotechnology has been a distinct scientific field for the past 10 years [6]. With its rapid development, it is difficult to cover all areas of this innovative science; however, it should be noted that many scientific fields in nanotechnology, such as engineering and science, are generally distinct from each other and can work together in the development of nanotechnology systems and devices [7]. At the nano-level, gravity is very low, electrostatic forces are inverted, and quantum effect occurs. In addition, the cells get reduced in size, the proportion of molecules on the surface increases compared to the interior, and this produces novel properties [8]. While current researchers in nanoscience and nanotechnology are exploring the features of this novel technology at the nanoscale, we can change the macro-features and produce significantly new materials and processes [9].

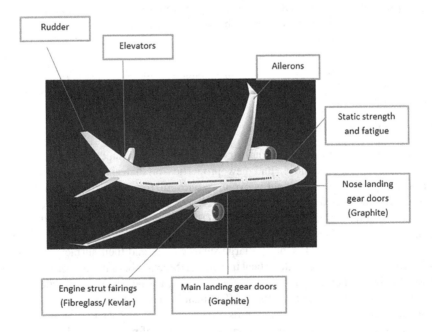

FIGURE 2.1 Nanotechnology in aerospace [5].

2.2 NANOMATERIAL STRUCTURE

The structure of carbon nanotubes is clearly defined as one or more coaxial cylindrical plates of graphite, usually measuring 100 and tens of nanometers in outer diameter, and ending with two semi-end domes [10]. The fullerene structure C_{60} can be found in the manufacture of carbon nanotubes. While the buckthorn is extended to form a stretched and slight tube about 1 nm (10^{-9} m) in diameter, it gives the uncomplicated shape of carbon nanotubes [11]. The basic element is graphite, which is formed one after another by the van der Waals forces, and it takes a two-dimensional coordinate structure throughout the sequence of processing to bend the planes of graphite is conceivable to generate a tube-shaped unified structure which is not made available in nature, commonly known as carbon nanotube [12]. In particular, two forms of carbon nanotubes are single-walled nanotubes (SWNTs), consisting of a single linear tubular unit, and multiwalled nanotubes (MWNTs) that are silica with respect to 0.34 nm apart distance which are made of tubes between different planes of graphite [13]. Therefore, carbon nanotubes can be seen as a graphite sheet wrapped in a tube unlike the diamond structure (sp_2 hybridization), and each carbon atom forms a 3D diamond cubic crystal structure, while graphite (sp_3 hybridization) is arranged like 2D sheet carbon atoms with each carbon atom possessing three closest structures [14]. The classification of nanomaterials based on dimensions is shown in Figure 2.2. A sheet of graphite is placed in the cylinders to form carbon nanotubes, and its properties rely upon the structural atomic arrangement, tube diameter, distance of the tubes, and the morphology or nanostructure [16]. The utilization of dissimilar production methods and precise developmental constraints, it is probable to acquire potential carbon nanotube morphology and properties for versatile applications in many scientific sectors of engineering [17].

2.3 SURFACE PROPERTIES OF NANOMATERIALS AND CNTs IN AEROSPACE

The size of the particles also has an excessive effect on their structural and mechanical properties, and there is no much difference in properties when a particle is at an unpacked state [18]. But, when a particle attains a size lesser than 100 nm, the properties of the nanoparticle change. The properties of a particle are resolved by the quantum size factors such as mechanical, magnetic, chemical, optical, thermal, and electrical [19]. These color changes are ascribed to alterations in their group type from connected to disconnected, owing to the isolated effect [20]. These effects of quantum in the nanoscaling are the primary reasons for tunability properties, and by merely tuning the size of the particle, we could do some modifications in their physical properties [21]. The functional properties of some nanomaterials are shown in Figure 2.3. A material's surface chemistry changes as the sample size is reduced; a particle's morphology, size, reactivity, surface area, surface charge, and potential must be considered during toxicity characterizations [23]. The last two characteristics, surface charge and potential, are of special importance because they control the stability of nanoparticles in solution [24,25]. General oxidative stress can also occur when nanomaterials enter the body. Carbon nanotubes (CNTs), discovered only 20 years ago, are perhaps the most widely known and used nanomaterials due to

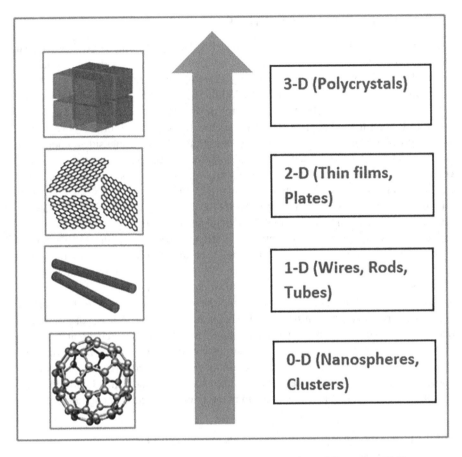

FIGURE 2.2 Arrangement of nanomaterials based on number of dimensions [15].

their carbon (C–C) bond interactions producing hexagonal lattice structures in which each atom bonds with three others, leaving an extra electron per atom to reinforce the bonds [26]. The CNTs gain much attraction in the aerospace industry as the carbon nanotubes along with the composite materials can be fastened to obtain the material considerably tougher, impervious to damage, and great strength on comparing to other advanced composites.

2.4 MICROSCOPY METHODS OF NANOMATERIALS

Nanomaterials play a major role, which are now a big industry. Improved synthesis processes and increased control and new properties of materials developed at the nanoscale indicate that this field is thriving. Although the prefix "nano" is included in nanometer materials, it is easy to see how micro-nanoscale materials occur. Most of these nano-sized particles can only be seen with an electron microscope such as transmission electron microscope (TEM), scanning electron microscope (SEM), scanning tunneling microscope (STM), and atomic energy microscope (AFM) [27].

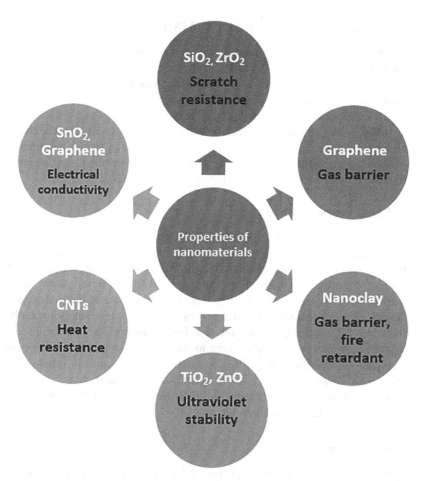

FIGURE 2.3 Properties of nanomaterials [22].

2.5 SCANNING TUNNELING MICROSCOPE (STM)

Scanning tunneling microscope (concept of quantum tunneling) operates by examining at the tip of an identical strident metal wire on the surface of a microscope. By allowing the tip at a very close distance to the surface of the microscope and concerning an electrical voltage to the tip or pattern, we can paint the shallow at a very low level to fix the individual molecules [28]. When a conductor tip is brought close to the surface for testing, a bias (voltage difference) applied between the two allows electrons to enter the tunnel through the vacuum between them. The resulting tunneling power is the purpose of flow tip location, pragmatic voltage, and local density of the states (LDOS). The information is obtained by scrutinizing the current as the tip position on the surface is scanned and usually displayed as an image. STM is an inspiring technology because it necessitates very spotless and firm surfaces, sharp tips, and first-rate and refined electronics, excellent vibration isolation and many enthusiasts tend to build their own microscopes [29].

2.6 ATOMIC FORCE MICROSCOPY (AFM)

Atomic force microscopy (AFM) is a surface scanning technique with subnanometer-scale resolution. The AFM technology designates a set of methods used for nondestructive surface studies at the nanoscale whose resolution is 103 times better than the resolution limit of optical microscopy. AFM is widely used at the nanoscale to collect data for topographic (surface) study along with various mechanical, functional, and electrical properties. It is commonly used to visualize the surface topography by recording the position of the sample relative to the tip and then the probe height, which allows for variation in mechanical, functional, and electrical properties along with continuous probe–pattern interaction [30].

2.6.1 Radar Absorbing Materials

The curiosity in radar absorbing materials (RAM) has extended to the business sector as they can be used to minimize electromagnetic interference due to the recent developments in electromagnetic devices entering the RF frequency range. Short carbon fiber-reinforced composites (CFRC) are ideally suited for the development of thin multipurpose RAM. Owing to their high execution of electrical and thermal qualities at comparatively low concentrations, composites dependent on polymers and carbon nanofillers have received significant interest in both academic and industrial societies. Extraordinary interest has been given to carbon nanostructure-filled polymer nanocomposites as electromagnetic absorbers in both military and civil applications in terms of their ability to modify electromagnetic and molecular properties at comparatively low quantities of nanofillers, and their lightweight, outstanding thermal tolerance, and high mechanical characteristics [31].

2.6.2 Nanochassis

One of the most ambitious priorities of the car industry is to incorporate light alloy bodywork. The latest cars will have been made heavier by introducing new mechanical parts and protective technology and by increasing comfort. To mitigate the fact, it has been possible to combine nanoparticles with less and light material to achieve the same mechanical resistance and lighter weight. This can vastly increase the characteristics such as strength, elasticity, and/or dimension flexibility, as well as unique characteristics such as indoor fire resistance and outdoor weather resistance. Another alternative offered is the plastic bodywork; along with the metal sections, they undergo electrostatic painting [32].

2.7 PROMINENCE OF NANOMATERIALS IN AEROSPACE INDUSTRY

Though nanomaterials are not extensively employed in aerospace manufacturing to date, their applications are predictable to be utilized in operationally complex compound panels within a few years. Carbon-based nanomaterials, including graphene, buckyballs, single- and multiwalled carbon nanotubes (SWCNTs and MWCNTs), carbon nanoparticles, and carbon nanofibers, are the most explored composite additives [33]. These nanomaterials can be combined with polymer matrix that is used in most composites and pragmatic along with the fluid during resin transfer molding.

After the curing process, they are embedded in the now solid polymer along with the standard macroscopic reinforcement materials, such as carbon fiber, fiberglass, Kevlar fiber, and honeycomb mesh [34]. The resulting composites are weightless and durable, with great mechanical, thermal, and electrical properties. It is expected that applications will include composite health monitoring and self-healing, greater aircraft brake disks that could disintegrate heat more proficiently, and strong interactive windscreens with dicing properties [35]. Figure 2.4 represents the application of nanomaterials in aerospace with environmental health and safety alarms. Nanomaterials can also be used for exploiting new technologies in the aircraft industry, such as increased safety and security, weight capacity, fuel catalysts, damping, bonding and curing of adhesives, lubrication, air filtering, jet engine block, communication and mobility, and reduced emissions and noise [36]. If the particles are electrically conductive, they can also improve the conductivity of the composite panel, allowing current to pass through the panel and into the surrounding structure, making it less vulnerable to damage from electrical discharge. Nanomaterials can also aid in electromagnetic shielding of sensitive components [37]. Aircraft engines are another subject for nanotechnology research. Some composites such as clay and ZrO associated with Y_2O_3 have exceptional heat resistance and can be used in the nacelles or exhaust ports of aircraft engines. Engine components can be coated in nanofilms that reduce friction and promote self-cleaning [38]. Nanosensors and strain gauges deployed in the interior of the engine and outer surface of the aircraft can give detailed readings in the regions of varying heat and pressure, providing a valuable feedback to monitoring systems. Composite manufacturing, even in the absence of nanomaterials, involves a hazardous work environment. Polymer composite matrix chemicals give off noxious fumes while they cure, and cleaning finished composites may involve acetone, methyl chloride, aliphatic amines, and methyl ethyl ketone (MEK) [39]. Exposure to these chemicals can lead to nausea, dizziness, vomiting, upper respiratory tract irritation, breathing difficulty, eye burns, and kidney and liver damage.

Machining fiber-reinforced composites releases micro- and nanofibers, which can be linked to lung cancer, asthma, mesothelioma, and pulmonary fibrosis. Those particles that are less than 6 μm in diameter are considered respirable, meaning that they can be taken into the lungs and other sensitive organs [40]. Those particles that are not inhaled can still be absorbed through the skin or eyes if they make contact, and careless transfer of materials containing these particles can lead to ingestion. In addition to its health consequences, microfiber dust is a fire hazard because particulate materials dispersed in air can explode. Among the nanotech applications, composite materials with carbon nanotube and graphene attachments have been regarded as promising prospects. In this review, an unbiased look at the progress of carbon nanocomposites has been designed to produce high-strength, low-density, high-conductivity nanoparticles. It provides an overview on the alternative approaches that can lead to potentially useful nanotube and graphene composites, highlighting the economic challenges that occur in the industry also, and this research work opens up the significant advances made in carbon nanocomposite over the past years and the discovery of new carbon nanocomposite processing technologies to improvise the functional impact of nanotube and graphene composites by providing a proper method of synthesis and improving the production of diverse composite based on carbon nanomaterials.

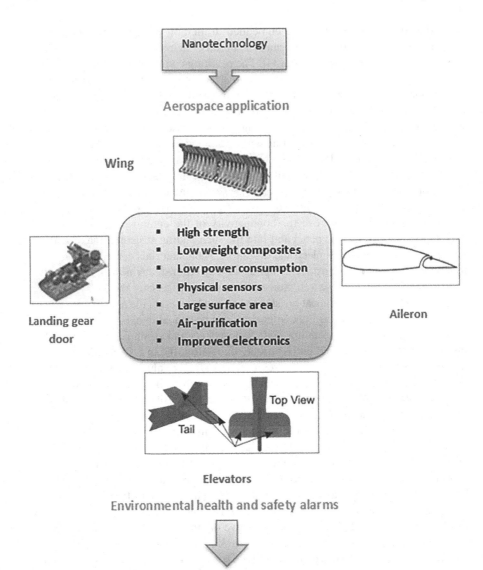

FIGURE 2.4 Nanomaterials in aerospace application and environmental health and safety alarms.

2.8 CNT STRUCTURES FOR AEROSPACE COMPONENTS

The special rigidity and tensile strength of carbon nanotubes (CNTs) make them suitable for use with polymer composites as reinforcement components. With the inclusion of carbon nanotubes, the strength and steadiness of a polymer substance can be significantly improved with limited weight increases. It may also improve a material's ability to withstand flame and vibration. However, in the last several years the price of nanotubes has plummeted drastically due to many attempts to realize mass processing of CNTs. This trend is anticipated to continue with refined nanotube synthesis techniques and additional production facilities. In a recent analysis article, four essential criteria were established for the successful fiber strengthening of composites: large aspect ratio, interfacial stress transfer, strong dispersion, and alignment. The carbon nanotubes usually have a very high aspect ratio. Many CNTs are on the order of a micrometer, while certain centimeters-long individual CNTs are synthesized. Because of their peculiar electric and structural features, carbon nanotubes are also not related to each other firmly. Consequently, a degree of interfacial stress propagation is limited to possible increases and enhancements of the mechanical characteristics of nanotube composites. A lot of research has been carried out to correct this issue by chemical functionality and carbon nanotube surface modification. For efficient and successful load transfers to nanotube, it is necessary to diffuse CNTs within the matrix. The influence of stress accumulation is decreased, and uniform stress distributions are decreased. The two key problems of dispersion are to isolate CNTs from each other and combine them with the polymer matrix equally. One of the most common means is to reinforce CNTs within a solvent. Shear mixing and magnetic stirring are also commonly used to mix nanotubes within a polymer. The matrix alignment of the CNTs can be the least important of all four nanotube composite criteria, since the criteria on alignment are mostly decided by the expected use. The most effective improvement in the fiber path will be the strongly directed fibers, but the cross-dimensions will not change to a minimum.

2.8.1 FUTURE SCOPE

In the automotive industry, nanotechnology plans a surge. This technology can have an important influence on the growing desire to optimize cars. Numerous future trends for smart cars will be identified in the range of nanotechnology. This is a progressive improvement in the features of new automobiles. There are electronically regulated sections, e.g., fuel injection, emission of exhausts, antilock brakes, automatic air-conditioning, headlight control, mechanical seat change, lateral control, and electronic hanging. Furthermore, it should be remembered that it is therefore better, as it includes a degree of artificial intelligence to compensate for driving mistakes. The car of the future will be connected to the other nearby vehicles and enhance the vision range. In the upcoming vehicle innovation, one of the fundamental capacities to remain universal competitive is nanotechnological skills. The development of nanotechnology would be all automotive subsystems. It requires the use of specialized nanoparticles in tires, reflective screen and mirror coatings, nanoparticle-enforced polymer and metals and adhesive primers, advanced technology in the fuel cell and storage of hydrogen, catalytic nanoparticles as a fuel additive, etc.

High-tech vehicles include headlights that automatically follow the street, and radar and heat sensors that recognize and assist people, animals, and objects on roads. This is a work of engineering, and manufactures are able to strip the assembly line from their state-of-the-art vehicles, to satisfy the increasing demand.

2.8.2 Cons of Nanomaterials

In addressing the pros and cons of nanotechnology, we also need to point out what the negative aspect of this technology can be seen as: The possible loss of jobs in traditional farming and manufacturing is included in the list of disadvantages of this technology and its development. Atomic weapons can still be more reachable and more harmful and influential. Nanotechnology can also make them more usable. Nanoscience also raised the health risk, as nanoparticles can cause inhalation complications due to their small size and a number of other deadly diseases can also quickly harm us by inhaling in the air for just 60 s. Nanotechnology is actually very costly and will cost you a lot of money to create. It is also very difficult to make, which is why nanotechnology goods are possibly costlier. The levels of life have been higher by nanotechnologies, but at the same time, pollution, like water contamination, has risen. Nanotechnology contamination is referred to as nano-pollution. For living creatures, this form of contamination is highly harmful. There is also little literature on the drawbacks of nanoparticles. There are only a handful of literature studies more focused on the distribution of medications. The formation of nanoparticles for drugs having a wide use as a detergent of polyvinyl alcohol that pose a toxicity problem. Nanoparticles have limited targeting capabilities, which is why it is not possible to discontinue the procedure And the cytotoxicity and alveolar inflammation indicate drug delivery with nanoparticles. The autonomic dysfunction condition by nanoparticles has a clear effect on the heart and vascular activity, particulate growth nanoparticles, unpredictable propensity to bubble, erratic polymeric transport mechanics, and often eruption of release. Researchers keep following the activities of nanoparticles without really understanding how their inventions could influence them. When technologies exceed human knowledge and understanding, underlying risks always exist. The ability to manipulate materials on a molecular basis is a great talent that may lead to abuse if left in the wrong hands. The possibility of a terrorist using this technology to produce lightweight, undetectable biological or nuclear weapons is especially alarming. The major concern is that these substances are engineered for one or more people to be potentially dangerous.

2.8.3 Conclusion

The nanocomposite or nanotechnology has a great potential in aerospace engineering which provides the outcome of material at high strength, weightless products, and resistance to corrosion with enhanced toughness and durability properties.

- The nanomaterials can be operated at a minimum maintenance and can be recycled again by making use of resources more efficiently, in turn increasing the productivity.

- In aerospace engineering, the nanoparticles, nanofibers, and nanofilms have enhanced electrical and thermal properties, cleaning, safer coating, resistant to corrosion, potential to toxicity facilities in various fields of aircraft components.
- The surface coating of aircraft parts protects from severe hazards more efficiently, and if not handled with proper care, it may be dangerous. The use of a nanomaterial structure in the manufacturing of airplanes makes it easy and convenient to repair.
- The operational cost is comparatively low and they can possess specific benefits and performance characteristics in comparison with the conventional metals and composites commonly used in the fabrication of various aerospace components.

REFERENCES

1. Gogotsi, Yury, ed. *Nanomaterials Handbook*. CRC Press, 2006.
2. Martin, Charles R. "Nanomaterials: A membrane-based synthetic approach." *Science* 266, no. 5193 (1994): 1961–1966.
3. Aruna, Singanahally T., and Alexander S. Mukasyan. "Combustion synthesis and nanomaterials." *Current Opinion in Solid State and Materials Science* 12, no. 3–4 (2008): 44–50.
4. Sharifi, Shahriar, Shahed Behzadi, Sophie Laurent, M. Laird Forrest, Pieter Stroeve, and Morteza Mahmoudi. "Toxicity of nanomaterials." *Chemical Society Reviews* 41, no. 6 (2012): 2323–2343.
5. Zhang, Hua. "Ultrathin two-dimensional nanomaterials." *ACS Nano* 9, no. 10 (2015): 9451–9469.
6. Martin, Charles R. "Membrane-based synthesis of nanomaterials." *Chemistry of Materials* 8, no. 8 (1996): 1739–1746.
7. Ulijn, Rein V., and Andrew M. Smith. "Designing peptide based nanomaterials." *Chemical Society Reviews* 37, no. 4 (2008): 664–675.
8. Khalajhedayati, Amirhossein, Zhiliang Pan, and Timothy J. Rupert. "Manipulating the interfacial structure of nanomaterials to achieve a unique combination of strength and ductility." *Nature Communications* 7, no. 1 (2016): 1–8.
9. Müller, Melanie, Alexander Paarmann, and Ralph Ernstorfer. "Femtosecond electrons probing currents and atomic structure in nanomaterials." *Nature Communications* 5 (2014): 5292.
10. Belyakova, O. A., Y. V. Zubavichus, I. S. Neretin, A. S. Golub, Yu N. Novikov, E. G. Mednikov, M. N. Vargaftik, I. I. Moiseev, and Yu L. Slovokhotov. "Atomic structure of nanomaterials: Combined X-ray diffraction and EXAFS studies." *Journal of Alloys and Compounds* 382, no. 1–2 (2004): 46–53.
11. Amendola, Vincenzo, and Moreno Meneghetti. "What controls the composition and the structure of nanomaterials generated by laser ablation in liquid solution?" *Physical Chemistry Chemical Physics* 15, no. 9 (2013): 3027–3046.
12. Moon, Robert J., Ashlie Martini, John Nairn, John Simonsen, and Jeff Youngblood. "Cellulose nanomaterials review: Structure, properties and nanocomposites." *Chemical Society Reviews* 40, no. 7 (2011): 3941–3994.
13. Batenburg, Kees Joost, Sara Bals, J. Sijbers, C. Kübel, P. A. Midgley, J. C. Hernandez, U. Kaiser, E. R. Encina, E. A. Coronado, and G. Van Tendeloo. "3D imaging of nanomaterials by discrete tomography." *Ultramicroscopy* 109, no. 6 (2009): 730–740.
14. Verma, Ayush, and Francesco Stellacci. "Effect of surface properties on nanoparticle-cell interactions." *Small* 6, no. 1 (2010): 12–21.

15. Kumar, Raghuvesh, and Munish Kumar. "Effect of size on cohesive energy, melting temperature and Debye temperature of nanomaterials." *Indian Journal of Pure and Applied Physics* 50, no. 5 (2012): 329–334.
16. Lundqvist, Martin, Johannes Stigler, Giuliano Elia, Iseult Lynch, Tommy Cedervall, and Kenneth A. Dawson. "Nanoparticle size and surface properties determine the protein corona with possible implications for biological impacts." *Proceedings of the National Academy of Sciences* 105, no. 38 (2008): 14265–14270.
17. Huang, Xiubing, Mu Yang, Ge Wang, and Xinxin Zhang. "Effect of surface properties of SBA-15 on confined Ag nanomaterials via double solvent technique." *Microporous and Mesoporous Materials* 144, no. 1–3 (2011): 171–175. DOI:10.1016/j.micromeso.2011.04.012.
18. Zhao, Feng, Jian Wang, Hongjuan Guo, Shaojun Liu, and Wei He. "The effects of surface properties of nanostructured bone repair materials on their performances." *Journal of Nanomaterials* vol. 2015, Article ID 893545, 11 pages, 2015. https://doi.org/10.1155/2015/893545.
19. Karakoti, A. S., L. L. Hench, and S. Seal. "The potential toxicity of nanomaterials—The role of surfaces." *JOM* 58, no. 7 (2006): 77–82.
20. Bera, Madhab, and Pradip K. Maji. "Effect of structural disparity of graphene-based materials on thermo-mechanical and surface properties of thermoplastic polyurethane nanocomposites." *Polymer* 119(2017): 118–133.
21. Holt, Martin, Ross Harder, Robert Winarski, and Volker Rose. "Nanoscale hard X-ray microscopy methods for materials studies." *Annual Review of Materials Research* 43(2013): 183–211.
22. Mishra, Raghvendra Kumar, Ajesh K. Zachariah, and Sabu Thomas. "Energy-dispersive X-ray spectroscopy techniques for nanomaterial." In *Microscopy Methods in Nanomaterials Characterization*, pp. 383–405. Elsevier, 2017.
23. Haynes, Holly, and Ramazan Asmatulu. "Nanotechnology safety in the aerospace industry." In *Nanotechnology Safety*, pp. 85–97. Elsevier, 2013.
24. Burgens, LaTashia. The Atomic Force Microscopic (AFM) Characterization of Nanomaterials. *Prairie View A and M Univ TX Coll of Engineering*, 2009.
25. Arepalli, Sivaram, and Padraig Moloney. "Engineered nanomaterials in aerospace." *MRS Bulletin* 40, no. 10 (2015): 804–811.
26. Shatkin, Jo Anne, Theodore H. Wegner, E.M. Ted Bilek, and John Cowie. "Market projections of cellulose nanomaterial-enabled products-Part 1: Applications." *TAPPI Journal* 13, no. 5 (2014): 9–16.
27. Boulos, Maher I. "New frontiers in thermal plasmas from space to nanomaterials." *Nucl. Eng. Technol* 44, no. 1 (2012): 1–8.
28. Lavrynenko, Sergiy, Athanasios G. Mamalis, and Edwin Gevorkyan. "Features of consolidation of nanoceramics for aerospace industry." In *Materials Science Forum*, vol. 915, pp. 179–184. Trans Tech Publications Ltd, 2018.
29. Aftab, S. M. A., Rabia Baby Shaikh, Bullo Saifullah, Mohd Zobir Hussein, and K. A. Ahmed. "Aerospace applications of graphene nanomaterials." In *AIP Conference Proceedings*, vol. 2083, no. 1, p. 030002. AIP Publishing LLC, 2019.
30. Anandan, S., Neha Hebalkar, B. V. Sarada, and Tata N. Rao. "Nanomanufacturing for Aerospace Applications." In *Aerospace Materials and Material Technologies*, pp. 85–101. Springer, Berlin, 2017.
31. Kumar, Indradeep, and C. Dhanasekaran. "Nanomaterial-based energy storage and supply system in aircraft." *Materials Today: Proceedings* 18(2019): 4341–4350.
32. Ghassan, Alsultan Abdulkareem, Nurul-Asikin Mijan, and Yun Hin Taufiq-Yap. "Nanomaterials: An overview of nanorods synthesis and optimization." In *Nanorods—An Overview from Synthesis to Emerging Device Applications.* IntechOpen, 2019, 1–24.

33. Mathew, Jinu, Josny Joy, and Soney C. George. "Potential applications of nanotechnology in transportation: A review." *Journal of King Saud University-Science* 31, no. 4 (2019): 586–594.
34. Zhang, Wei, Seiji Yamashita, and Hideki Kita. "Progress in tribological research of SiC ceramics in unlubricated sliding—A review." *Materials & Design* 190 (2020): 108528.
35. Al-Jothery, H. K. M., T. M. B. Albarody, P. S. M. Yusoff, M. A. Abdullah, and A. R. Hussein. "A review of ultra-high temperature materials for thermal protection system." In *IOP Conference Series: Materials Science and Engineering*, vol. 863, no. 1, p. 012003. IOP Publishing, 2020.
36. Dobrovolskaia, Marina A., and Scott E. McNeil. "Immunological properties of engineered nanomaterials." *Nature Nanotechnology* 2, no. 8 (2007): 469.
37. Batenburg, Kees Joost, Sara Bals, J. Sijbers, C. Kübel, P. A. Midgley, J. C. Hernandez, U. Kaiser, E. R. Encina, E. A. Coronado, and G. Van Tendeloo. "3D imaging of nanomaterials by discrete tomography." *Ultramicroscopy* 109, no. 6 (2009): 730–740.
38. Haynes, Holly, and Ramazan Asmatulu. "Nanotechnology safety in the aerospace industry." In *Nanotechnology Safety*, pp. 85–97. Elsevier, 2013.
39. Barako, Michael T., Vincent Gambin, and Jesse Tice. "Integrated nanomaterials for extreme thermal management: A perspective for aerospace applications." *Nanotechnology* 29, no. 15 (2018): 154003.
40. Abbasi, Sadaf, M. H. Peerzada, Sabzoi Nizamuddin, and Nabisab Mujawar Mubarak. "Functionalized nanomaterials for the aerospace, vehicle, and sports industries." In *Handbook of Functionalized Nanomaterials for Industrial Applications*, pp. 795–825. Elsevier, 2020.

3 Lightweight Polymer–Nanoparticle-Based Composites
An Overview

Harrison Shagwira and F.M. Mwema
Dedan Kimathi University of Technology

Thomas O. Mbuya
University of Nairobi

CONTENTS

3.1 Introduction 32
3.2 Classification of Composite Materials..32
    3.2.1 Metal Matrix Composites...32
    3.2.2 Ceramic Composites...33
    3.2.3 Polymer Composites 33
3.3 Application of Polymer Composites 34
    3.3.1 Aerospace Industry 34
    3.3.2 Automotive Industry 34
    3.3.3 Marine Industry 35
    3.3.4 Microelectronics 35
    3.3.5 Medical Applications..36
    3.3.6 Construction Industry 36
3.4 Processing of Polymer-Based Composites 37
    3.4.1 Autoclave Molding..37
    3.4.2 Out-of-Autoclave Quickstep Molding 38
    3.4.3 Liquid Molding 38
    3.4.4 Filament Winding Process..40
3.5 State-of-the-Art Review of Polymer–Nanoparticle Composites 40
    3.5.1 Micro-composites: Sand–Plastic Composite......................................41
    3.5.2 Nanocomposites: Plastic–Silica Nanoparticle Composites 42
3.6 Conclusion 45
References..46

3.1 INTRODUCTION

A lot of research has been undergoing for a long time in coming up with composite materials. Different composites have different applications depending on their properties. It is, therefore, necessary to have a review of the literature to understand the existing gap especially in the construction industry. This chapter includes an overview of classification, applications, and properties of polymer-based composites, and theory of processing and manufacturing of polymer-based composites and products. Most importantly, it includes a detailed review of the research and progress on the polymer-based composites with an emphasis on polymer–silica nanoparticle-based composites.

3.2 CLASSIFICATION OF COMPOSITE MATERIALS

Composite materials refer to the class of materials that consist of two or more components that are diverse in their physical and chemical properties (Altenbach, Altenbach, & Kissing, 2018). The combination of two or more materials eventually results in a new material with different properties from the single composition (Nielsen, 2005). Individual materials used in the production of composite material components are discrete and separable within the final composite material configuration; however, composites have to strictly be differentiated from mixtures and solution of solids (Koniuszewska & Kaczmar, 2016). Studies carried out have documented many beneficial features of composite materials. First, they are stronger (Murr, 2014), and secondly, they are generally of lower density and less expensive compared to the original materials (Dawoud & Saleh, 2019). Composite materials are divided into three categories based on the matrix constituents: metal matrix composites, ceramic matrix composites, and polymer matrix composites. Composites are also classified according to the size and shape of the reinforcing material structure, for example, particulate or fibrous reinforced composite.

3.2.1 METAL MATRIX COMPOSITES

These are metals reinforced with other metals, organic compounds, or ceramic compounds (Casati & Vedani, 2014). The making of these composites involves the dispersion of reinforcement in the metal matrix basically to improve the properties of the base materials. The study carried out by Ramnath et al. (2014) on aluminum metal matrix composite reported an enhancement in terms of strength, strain, hardness, wear, and fatigue of the aluminum metal. However, the study reported a decline in tensile strength. A study conducted on reduced graphene oxide–metal composite for application in water purification (Sreeprasad, Maliyekkal, Lisha, & Pradeep, 2011) found that the graphene-based composite was efficient for the purification process. Besides, the study noted that the composites can also be used in a wide range of applications such as in catalysis and fuel cells. Another study (Casati & Vedani, 2014) on metal matrix composites enforced by nanocomposites showed remarkable results such as increased hardness, mechanical strength, wear resistance, creep behavior, and damping properties (Macke, Schultz, & Rohatgi, 2012). Although the composites were said to aid in the reduction of costs incurred on conventional monolithic alloys, some aspects such as clustering of particles, the complexity involved in the fabrication process, and clarity on the reactions between ceramic nanoparticles or carbon nanotubes would require improvement and further research.

3.2.2 CERAMIC COMPOSITES

They consist of a ceramic matrix combined with a ceramic (oxides, carbides) dispersed phase (Porwal et al., 2013). They are particularly designed to enhance the toughness of conventional ceramic materials, which are naturally brittle (Walker, Marotto, Rafiee, Koratkar, & Corral, 2011). There are various studies that have been conducted on fiber ceramic composites. In the previous years, the focus was mainly on carbon nanotube (CNT)-reinforced glass and ceramic composites. The authors have documented enhanced properties such as toughness, strength, and electrical conductivity over original ceramic (Choi et al., 2018; Katoh & Nakagama, 2014). Similarly, Porwal et al. (2013) reported a significant enhancement in mechanical, electrical, and thermal properties under graphene ceramic matrix enhancement. Additionally, Walker et al. (2011) found that graphene ceramic composite has the potential to improve the mechanical properties of polymers.

3.2.3 POLYMER COMPOSITES

Polymers fall into two categories: thermoplastic and thermosetting (Altenbach et al., 2018). The most commonly used thermoplastic materials include polypropylene, polyethene, and polyvinyl chloride, while epoxy, polyester, and phenolic are the mostly used thermosetting matrices (Nielsen, 2005). Recently, natural fibers as polymer composite fillers have gained much interest due to their better physical and mechanical properties as compared to synthetic fibers, e.g., glass. Some of these natural fibers include hemp, sisal, jute, and flax, among others (Pickering, Efendy, & Le, 2016). Their advantages over conventional carbon and glass fibers include non-abrasiveness, low density, better tensile properties, low cost, and reduced health risk, among others (Sreeprasad et al., 2011). The main applications of polymer composites fall under construction, packaging, aerospace, automotive, and sports industries (Murr, 2014). Despite these advantages, these composites are limited by the incompatibility that exists between the hydrophobic thermoplastic and hydrophilic natural fiber matrices. There is significant research that has been undertaken on polymer–fiber composite focusing on specific applications, enhancements of properties, and optimization of performance. The use of natural fibers in polymers results in materials that are eco-friendly, less expensive, and excellent in tensile behavior and can be used as an alternative to conventional fibers such as glass (Wang et al., 2011). However, the limitation of these (polymer–fiber) materials is that the strength of the polymer is dependent on the fiber loading. This may not be advantageous since the increment in fiber weight results in decreased tensile strength.

A polymer–nanoparticle composite material is produced by incorporating synthetic or natural nanoparticles into a polymeric matrix. Silica nanoparticles may be obtained from sources of natural silica such as sand, clay, and quarry dust. The inclusion of silica nanoparticles into a polymeric matrix can improve the thermal, mechanical, and fire-retardant properties of the polymer material. A polymer–nanoparticle composite is a material containing one of its phases (reinforcing material) in a nanometer-sized structure, and it is considered to be a nanocomposite; otherwise, the composite material is a micro-composite. It is worth emphasizing that the main features of polymer–silica nanoparticle composite material have close relationships with each phase's physical and chemical characteristic properties and also with the size of silica nanoparticles and their interfacial adhesion between the matrix and silica nanoparticles.

Nanocomposites are composites that contain one of the phases in nano-size (10^{-9} m). These composite materials started to be produced because of their superior physical, thermal, and mechanical properties in comparison with traditional composites and micro-composites. Besides, the preparation techniques and processing of these nanocomposites show different challenges as a result of the stoichiometry in the nano-phased and elementary structure. Nano-phased filler materials are integrated into the matrix of the composite to enhance the properties of the nanocomposites (Asmatulu, Khan, Reddy, & Ceylan, 2015). Polymer nanocomposites are an interesting material category that exhibits distinct physical and chemical properties that cannot be achieved by individual components. Due to their exciting capabilities in many applications in environmental sustainability and addressing various environmental challenges, polymer nanocomposites have increasingly attracted thorough research attentions (Chowdhury, Amin, Haque, & Rahman, 2018). Some of the polymer nanocomposites include PLA/fumed silica/clay (PLA-fsi-clay) nanocomposites, PVA/silica/clay (PVA-si-clay) nanocomposites, PVA/fumed silica/clay (PVA-fsi-clay) nanocomposites, PF/fumed silica/clay (PF-fsi-clay) nanocomposites, and ST-co-GM/fumed silica/clay (ST-co-GMA-fsi-clay) nanocomposites (Rahman, Chang Hui, & Hamdan, 2018).

3.3 APPLICATION OF POLYMER COMPOSITES

3.3.1 Aerospace Industry

In order to improve the fuel economy, carrying capacity, and maneuverability of airplanes, there is the need to adopt the use of new materials that are low in weight and high in strength (Zhang, Chen, Li, Tian, & Liu, 2018). Composites materials exhibit these properties and are therefore attractive for aerospace applications. For instance, American Airlines, which constitutes a fleet of 600 planes, could immensely save on the fuel cost by reducing the aircraft weight (Morris, 2018). Several airplane manufacturers have opted for the application of natural fiber-based composites to minimize the cost of production and enhance the use of eco-friendly materials. Most of the components in aircraft are currently manufactured from polymer composites. These include aircraft body, wings, fuselage, doors, tail, and interior that are mostly manufactured from carbon fiber-reinforced plastic (CFRP) due to its high strength-to-weight ratio (Irving & Soutis, 2015). Sections of the wings and tail are manufactured from fiberglass (Maria, 2013).

3.3.2 Automotive Industry

There is pressure to manufacture light, fuel-efficient, low-cost, and green automobiles in modern society. The use of polymer-based composites on some components of an automobile has been shown to enhance the low-weight and green manufacturing of cars (Witik, Payet, Michaud, Ludwig, & Månson, 2011). Some of the components in the automotive industry are extensively manufactured from polymer-composite bumper beams, battery boxes, and seatbacks produced from glass mat thermoplastics (GMT) (Witik et al., 2011); interior headliner, engine cover, underbody system, air intake manifold, deck lid, instrument panel, bumper beam, front-end module, load floor, air duct, airbag housing, and air cleaner housing produced from glass-reinforced plastics (GFRP) (Friedrich & Almajid, 2013; Holbery & Houston, 2006); roof, rear spoiler, trunk lid, side panels, floor panel, hood frame, chassis/monocoque, tailgate, hood, bumper, and fender produced from carbon fiber-reinforced composites (CFRP) (Mitschang & Hildebrandt, 2012).

3.3.3 Marine Industry

Polymer composite materials have found a great application in building marine structures. This is attributed to better physical, mechanical, chemical, and thermal properties these composites possess. Some of the desirable properties include low weight, good long-term properties (no corrosion), and the ability to produce components with complex shapes with affordable tooling. The low weight of marine construction of the ship is important for low fuel consumption and effective performance. For instance, the speedboat revolver 42, which is a result of the collaborative work of Michael Peters Yacht Design and the Milan-based studio, is a remarkable example of the application of polymer composites (Neşer, 2017). The hull and deck are made from cystic vinyl ester resin and a core cell M-foam and enhanced with carbon fibers (Koniuszewska & Kaczmar, 2016). Despite the ship having a large mass, it can accelerate up to a speed of 68 knots. For a similar reason, sailboat wings are primarily fabricated from carbon spar. Additionally, there has been a lot of research aiming at improving the properties of polymer composite materials for the underwater application while prolonging their underwater life. The aim is to replace the old traditional glass composite with thermoplastic matrix composite for large submarine elements (Neşer, 2017).

3.3.4 Microelectronics

The electronics industry is growing rapidly, and polymer composite materials are increasingly finding great electronic applications. This is because of some desirable properties such as low thermal expansion, low/high electrical conductivity, low dielectric constant, high thermal conductivity, and/or electromagnetic interference (EMI) shielding effectiveness that is required in electronics applications. Some of the applications of polymer composites in microelectronics include thermal interface materials, photovoltaic device, interconnections, organic light-emitting diode (OLED), housings, actuator, sensors, connectors, substrates, encapsulations, electroluminescent device, heat sinks, printed circuit boards, die attach, lids, interlayer dielectrics, displays, electrodes, batteries, and electrical contacts (Wei, Hua, & Xiong, 2018). For example, the use of carbon fillers such as graphene, fullerene, and carbon nanotubes (CNTs) in the polymeric matrix has proven to be appropriate for the detection of various kinds of molecules, e.g., gases, heat, biomolecules, temperature, and pH (Rahaman, Khastgir, & Aldalbahi, 2019).

3.3.5 MEDICAL APPLICATIONS

Polymer-reinforced composites play an important part in the science of polymers because of their typical properties, e.g., solvent resistance, strong viscoelastic properties, stability at high temperatures, and high mechanical strength. Thermosetting polymers cannot be melted or reshaped once they have been produced, and as a result, they are found to be suitable in various applications that require these properties. Polymer composites show biodegradability, high cell adhesion, low inflammatory response, and biocompatibility when implanted for applications of tissue engineering (Ramakrishna, Mayer, Wintermantel, & Leong, 2001). They find great application in biomedical fields, such as in replacement of hardened tissue, preparation of dental materials, wound dressing, polymeric heart valves, medical devices such as electrocardiographs, bone formation, and prosthetic sockets (Zafar et al., 2016). For example, polyolefins cross-linked with poly(styrene-block-isobutylene-block-styrene) can be used as a heart valve (Madhav, Singh, & Jaiswar, 2019). Properties of polymer matrices can be modified by the addition of metal fillers; e.g., GO-modified epoxy polymer matrix displays an increase in mechanical and thermal properties, whereas Ag nanoparticles enhance dielectric and antimicrobial properties (Qi et al., 2014). Polymer composites, however, have certain disadvantages including poor cell affinity and the release of acidic by-products (Zagho, Hussein, & Elzatahry, 2018).

3.3.6 CONSTRUCTION INDUSTRY

In the past 30 years, innovative polymer composites have become appealing in the construction industry as new structural materials and there has been an increase in their usage in the reconstruction of existing bridges and buildings. The research and development strategies for polymer composites for application in the construction industry are continuously underway, and the advancement accomplished on this exciting material has continued to increase to satisfy the construction industry demands. Some of the applications include the use of composites in the rehabilitation and repair of wood, steel, concrete, and masonry structures and all-composite applications in constructing structures, which include the construction of bridges and buildings (Medina et al., 2018). For example, the usage of fiber-reinforced polymer composites (FRPC) increases energy absorption efficiency and load-carrying capability of slabs made up of FRPC. Generally, stress transmission throughout the crack improves by self-strengthening, which slows the formation of cracks, and therefore, FRPC reinforcement is capable of achieving its entire capability to strengthen the slabs (Mosallam, 2014). Additionally, pultruded fiber-reinforced polymer (PFRP) composites exhibit the electromagnetic transparency and radio wave reflection properties. These non-magnetic properties of PFRP composites are desirable in applications requiring construction of facilities containing delicate instrumentation (Alberto, 2013; Gand, Chan, & Mottram, 2013). However, not much has been reported on the applicability of polymer–silica in the construction industry.

3.4 PROCESSING OF POLYMER-BASED COMPOSITES

The processing of polymer-based composites involves two major steps: melting and forming in a mold (die). For composites with thermoplastic matrix, the consolidation process is achieved by cooling; on the other hand, for thermoset matrix, consolidation is achieved by curing (Baran, Cinar, Ersoy, Akkerman, & Hattel, 2017). Concerning the thermoset matrix composite, the curing can be conducted at room temperature although it can be quickened through the application of heat typically through an oven in vacuum conditions (Singh, Chauhan, Mozafari, & Hiran, 2016). Notably, curing, which enhances successful cross-linking and polymerization process of the hydrocarbon chains, can be enhanced by other forms of energy, besides the heat, and these may include X-ray, electron beam, ultraviolet, and microwave curing (Abliz et al., 2013).

There are various methods used in the processing of polymer composites, and some of the common ones include autoclave molding, out-of-autoclave Quickstep molding, liquid molding, and filament winding processing.

3.4.1 AUTOCLAVE MOLDING

Autoclave molding is among the open molding techniques in which vacuum, pressure, and heat of the inert gases are used to cure the molded component. Figure 3.1 shows a schematic diagram showing the autoclave molding process setup. In terms of operation, the molded component is put in a plastic bag containing a vacuum created by a vacuum pump. The presence of a vacuum prevents the molded component from

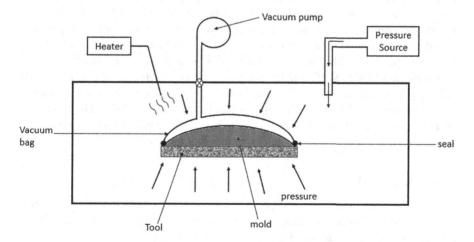

FIGURE 3.1 Schematic diagram showing the autoclave molding process setup.

coming into contact with volatile materials and air inclusions. Then, curing and densification of the part are done by the application of heat and pressure of the inert gas in the autoclave. Eventually, autoclave curing allows the production of uniform homogeneous components due to heating from both sides (Ouarhim, Zari, Bouhfid, & Qaiss, 2018). According to Ghori, Siakeng, Rasheed, Saba, & Jawaid (2018), curing times can be automated using the controller according to a specific cure profile to pressurize and heat the unconsolidated laminate stack. Before placing a vacuum bag above the entire assembly of tools, layers of breather and release film are first put in place (Alagirusamy, 2010). The sealed air is drawn out of the assembly basically to pressurize it for maximum fiber reinforcement and minimum creation of voids in the cured composite part, requiring minimal finishing. The technique is comparatively expensive, and it is applied in the production of high-quality aerospace components. Additionally, this method has some benefits; for example, the applied pressure helps in binding the composite materials by increasing the density of the lining and strength of the bond making them more compact, the ability to produce composites with high-fiber load, and the production of high-quality components.

3.4.2 Out-of-Autoclave Quickstep Molding

The applications of out-of-autoclave technique have increased in popularity over the last decade due to the ability to cure autoclave-quality materials/components in vacuum-bag-only (VBO). To achieve high-dimensional tolerance and low porosity, VBO prepregs rely on particular processing techniques and microstructural features. The Quickstep technique of producing components made of fiber-reinforced composites depends on the conduction heating principle. During processing, glycol-based heat transfer fluid (HTF) is used to transfer heat and pressure to the processed component that is uncured. According to Drakonakis, Seferis, & Doumanidis (2013), the high thermal conductivity and heat capacity of HTF as compared to those of the air enable the processing temperature to be controlled effectively than in an autoclave or oven. It uses a conventional layup sealed in vacuum bag processed in a pressure chamber that has the HTF (Khan, Khan, & Ahmed, 2017). Afterward, the processed component is sandwiched between two flexible membranes in the pressure chamber by which the HTF supplies the necessary heat and pressure to consolidate and cure the matrix–fiber interface. Under this process, the temperature is maintained by circulating the HTF in the pressure chamber, thereby enhancing the rapid cooling and heating rates and the control of resin viscosity in relation to the findings of Hernández, Sket, Molina-Aldareguıa, González, & LLorca (2011). The quick heat energy transfer into the curing fiber is the main technique in this process. Figure 3.2 shows a typical out-of-autoclave Quickstep molding process.

3.4.3 Liquid Molding

Liquid composite molding (LCM) comprises several composite production methods, for example, Seemann composite resin infusion molding process (SCRIMP), resin transfer molding (RTM), injection compression molding (ICM), and vacuum-assisted RTM (VARTM) processes. Such a technique can produce complex-shaped,

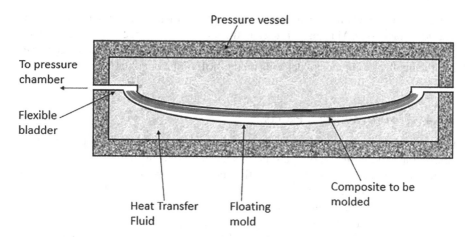

FIGURE 3.2 Schematic diagram showing the out-of-autoclave Quickstep molding process (Ogale & Schlimbach, 2011).

high-quality fiber-reinforced composite and is therefore predominantly used in the automotive, aerospace, civil, and marine industries. It is the most commonly used processing technique for polymer matrix, and this is because of its low cost (Finkbeiner, 2013). RTM is the primary method that has given rise to most of the variations. Figure 3.3 demonstrates a flow diagram of the steps of a typical RTM method. The preform is first formed and put in the mold compartment. Once the mold is closed, a polymeric resin is introduced into the mold chamber, which saturates the preform and ejects the existing air in the mold chamber (Walbran, Bickerton, & Kelly, 2013). A curing process is triggered causing the cross-linking of the thermoset resin, either during or after mold injection, to produce a solid piece.

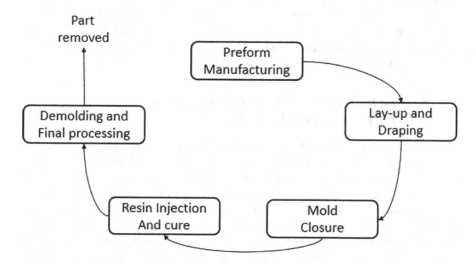

FIGURE 3.3 Flow diagram of the steps of a typical RTM method.

After the component has been cured enough, it can then be taken out of the mold. The advantage of LCM is that it is used in the production of larger and more complex parts and has short cycle times and rejection rates. The process is labor-intensive, and the quality of composite depends on the skill of the operator (Hamidi & Altan, 2017).

3.4.4 Filament Winding Process

Filament winding is a method by which composite components are produced by continuously winding fibers on a specially oriented rotating mandrel. This method of polymer composite production is the most economical in the processing of symmetric composite components in areas requiring mass production (Abdalla et al., 2007). This technique is mainly used with circular or oval hollow, sectional parts, e.g., tanks and pipes. This offers a wide range of applications, from small hollow products such as gas cylinders to huge products such as cryogenic tanks. The fibers are first passed through a bath of resin where the fibers are moistened before they are winded. The desired properties of the composite components can be achieved by varying the winding thicknesses and the number of layers. The fibers are supplied with enough tension to compact them on the mandrel. Varying the revolution of the mandrel and movement of the carriage produces a variety of winding patterns. This entire process normally is carried out at room or high temperatures. After curing, the mandrel is removed from the produced composite part for reuse (Mack & Schledjewski, 2012). The advantages of filament winding include excellent mechanical properties achieved by the use of constant fibers, process speed, better fiber and material control, good interior finish, good thickness control, and high intensity of the strengthening procedure. The main disadvantages of this technique are as follows: the difficulty of winding complex profiles which may require the use of complex equipment, limitations on convex-shaped components, poor external finish, low-viscosity resins, and high cost of the mandrel. Some of the applications of filament winding include the production of open-ended products, such as gas cylinders and tube systems, and closed-end products, such as chemical tanks and pressure vessels (Minsch, Herrmann, Gereke, Nocke, & Cherif, 2017).

3.5 STATE-OF-THE-ART REVIEW OF POLYMER–NANOPARTICLE COMPOSITES

An extensive literature on studies involving plastic–nanoparticle composites is available. The focus of such studies has been on the enhancement of properties and optimization of the performance of the composites. The emphasis has been on the polymer matrix–reinforcement adhesion, strength, and matrix–reinforcement ratios. A critical review of the literature review revealed that polymer-based composites can be classified based on the size of the reinforcement particles with an emphasis on mixture ratios. This is the focus of this research. As such, the review in this subtopic is analyzed under micro- and nanoparticle composites for comparison purposes.

3.5.1 Micro-composites: Sand–Plastic Composite

In polymer–sand composites, the effect of the ratio of matrix to reinforcing sand particles plays an important role in the properties and performance of the composites. For instance, Sultana et al. (2013) investigated sand–polyester composite prepared at varying weights of 10, 20, 30, 40, 50, and 60 wt.% of sand. The tests conducted were water absorption rate, comprehensive strength, flexural strength, hardness, and thermal conductivity tests. It was found that the percentage of water absorbed increased as the time of immersion increased. Additionally, the water absorption rate increased with an increase in sand composition. A similar result was reported by Bajuri, Mazlan, & Ishak (2018) who investigated the effect of micro-silica concentration on epoxy. The water absorption rate for the composite increased with an increase in the concentration of micro-silica particle due to the hydrophilic nature of silica. Moreover, the compressive strength and the flexural strength of the composite decreased with an increase in the amount of sand. Similar results were reported by Herrera-Franco, Valadez-Gonzalez, & Cervantes-Uc (1997) where the flexural and tensile properties of the HDPE–sand composite decreased as the amount of silica sand was increased beyond 30% wt. composition. This was attributed to the poor adhesion between the polymer and the silica sand interfaces. However, high flexural strengths were exhibited, indicating that the composite was brittle as a result of poor adhesion between the matrix and the fiber. When the Vickers hardness test was conducted, it was noted that the hardness of the composites increased with an increase in sand composition. The thermal conductivity decreased with an increase in the sand content.

In another study, Seghiri, Boutoutaou, Kriker, & Hachani (2017) mixed sand dune and r-HDPE to form sand dune–plastic composite in 30, 40, 50, 60, 70, and 80 wt.% HDPE. The tests that were conducted included flexural rigidity test, impermeability test, and density test. It was found that the density varied from 1.379 to 1.873 g/cm^3. The composite exhibited good impermeability as compared to clay tile. Additionally, the flexural rigidity of all the composite mixes was lower than that of tile made up of clay. The results are comparable to those of the studies conducted by Bajuri et al. (2018) in which silica particles were used as filler material to enhance the properties of composite reinforced with kenaf. For 10 min, the silica particles were deposited into the epoxy matrix using a homogenizer at a speed of 3,000 rpm before being injected into the fibers. It was found that the addition of silica particles typically decreased the mechanical properties of the composite. However, better mechanical properties were achieved with the addition of 2 vol.% of silica with the flexural strength, flexural modulus, compressive strength, and compressive modulus of 43.8 MPa, 3.05 GPa, 40 MPa, and 1.15 GPa, respectively.

Similarly, Aghazadeh Mohandesi, Refahi, Sadeghi Meresht, & Berenji (2011) produced polyethene terephthalate (PET)–sand composite by blending recycled PET waste plastic in the form of molten polymer with silica sand particles at 5%–40% sand particle weight concentrations. The average diameter of sand particles in the form of particulate composites ranged from 0.062 to 0.35 mm. The produced composites were tested with three bending points and compression at varying temperatures of 20°C, 25°C, 40°C, 60°C, and 80°C. For comparison purposes, the related compression strength analyses were numerically modeled to approximate the cohesive strength between the fine particles defining the structure of the composite materials evaluated. The results indicated that the tested composites exhibited the maximum pseudo-cohesive strength and mechanical strength at 25°C. Further, composite compressive strength increased with an increase in the percentage weight of the sand particles by up to 10%, and it decreased with further increase in percentage weight (i.e., 20% and 40% sand).

Additionally, Abdel-Rahman, Younes, & Yassene (2018) investigated the effect of varying composition of clay (silica) in 0%, 3%, 5%, and 10% weight concentrations in the unsaturated polyester polymer matrix and varying gamma irradiation at 30 and 50 kGγ on the polyester–clay composite's mechanical properties. The results revealed that there was an improvement in the compressive strength as the composition of the clay content in unsaturated polyester matrix was increased up to 5 wt.%. These observations under the influence of γ-irradiation were attributed to the adhesion between and unsaturated polyester matrix and clay additive within the composite structure. The results obtained in TGA also showed that the composite's thermal stability improved as the composition of clay increased. The use of 50 kGγ radiation showed good thermal stability as compared to the use of 30 kGγ radiation.

The chemistry of the matrix material has also been shown to influence the characteristics of polymer–sand composites. In a study, Slieptsova, Savchenko, Sova, & Slieptsov (2016) produced plastic–sand composite using recycled plastic and sand as a reinforcement additive. A comparison based on characterization techniques was made between the composites made from polyolefin (low-density polyethene (LLDPE) and polyethylene (PE)) and polyester (PET) matrices. Various composition alteration methods were used, such as the addition of compatibilizers, filler surface treatment, and the production of polymer blends. The effect of varying the composite constituents in its structure on the mechanical composite properties was determined. The produced composite comprising PET–polycarbonate (PC) mixture displayed improved thermal and mechanical properties than the composites comprising polyolefin. These composites find application where there is a need to use thinner and lighter materials having excellent thermal stability and high rigidity.

3.5.2 NANOCOMPOSITES: PLASTIC–SILICA NANOPARTICLE COMPOSITES

A lot of research exists on the influence of the reinforcement ratios of silica nanoparticles on the plastic composites. In a study by Ahmed & Mamat (2011), HDPE–silica nanoparticle composite was produced containing silica nanoparticles with an average size of silica nanoparticle of less than 100 nm. The silica nanoparticles were generated in several steps in ball mills and combinations of heating. The HDPE–silica nanoparticle composite was produced by mixing HDPE with silica nanoparticles while varying the concentration of silica nanoparticles by 5, 10, 15, and 20 wt.% and then through compression molding. The nanocomposite was evaluated based on the physical properties, thermal properties, mechanical properties, and microstructure. It was found that there was an improvement in the physical properties with the addition of silica sand nanoparticles as a reinforcement additive in HDPE. Additionally, there was an improvement in the mechanical properties of HDPE–silica nanoparticle composite with an increase in silica nanoparticles with an optimum value of 15 wt.%. DSC results showed that the crystallinity of silica sand nanoparticles decreased.

In a research conducted by Krasucka, Stefaniak, Kierys, & Goworek (2015), cross-linked copolymer resins were used in different chemical compositions to prepare silica gel and polymer–silica nanoparticle composite. In order to study the formation of porosity in different polymer templates, the structure of calcined pellets and nanocomposite pores was studied using the typical adsorption process. Based on their hydrophobicity and geometric structure, examination of the porosity parameters describing the studied materials within meso- and micropores displayed a varying degree of silica nanoparticle portion penetration into the polymer matrix. It has been reported that the adsorption–desorption process can be affected by the pore blockage and cavitation effects (Reichenbach, Kalies, Enke, & Klank, 2011; Thommes, 2010). The addition of silica nanoparticles in the polymer matrix allows swelling in solvents to be eradicated and the entire matrix structure to be reinforced. Therefore, the interaction between the polymer matrix and an inorganic additive plays an important role in the formation of pores inside a composite which correlates with the swelling of a polymer composite. A difference in the polymer structure leads to a different interaction between the polymer and an inorganic additive. Ji et al. (2003) conducted a similar study by reacting tetraethoxysilane (TEOS) with propyl methacrylate to form a silica–polymer nanocomposite. The results indicated an improvement in thermal and mechanical properties. Therefore, the addition of silica nanoparticles in the polymer matrix significantly changes the porosity and morphology of the original particles as well as influencing the mechanical and thermal properties of the polymer.

Similarly, Peng & Kong (2007) prepared a nanocomposite made up of polyvinyl alcohol/silica nanoparticles using a combination of two methods: self-assembly and solution compounding methods. The findings indicated that the intense interaction of the particle with the matrix is completely inhibited and that the homogeneity of spherical silica nanoparticle dispersion in the polyvinyl alcohol (PVA) matrix is achieved. Atomic force microscopy height profiles showed heterogeneity in the surface of the nanocomposite which is influenced by the concentration of silica nanoparticles. The values of roughness assessed indicated that an increase in silica nanoparticle content leads to a rougher surface. Therefore, structural changes occur with an increase in silica nanoparticles into the composite polymer matrix.

Moreover, Younes et al. (2019) studied the effect the amount and type of binder have on the thermal and porous properties in a silica gel composite. A selection of four binders was made: polyvinylpyrrolidone (PVP), polyvinyl alcohol (PVA), gelatin, and hydroxyethylcellulose (HEC). It was found that silica gel powder (SGP) containing 2 wt.% of PVP composite demonstrated improved thermal and porous properties. Higher thermal conductivity of 32% more than that of SGP was found for PVP 2 wt.% composite. By comparison, the adsorption of water uptake for both SGP and PVP 2 wt.% of the composite remained the same, while there was a 12.5% increase in the volumetric uptake for the composite. The composites evaluated were considered to be appropriate for high-performance adsorption cooling system designs.

Furthermore, Kalambettu, Venkatesan, & Dharmalingam (2012) produced polymethacrylate–silica composite membrane and polyvinyl alcohol–silica composite membranes using the sol–gel method and studied their suitability in medical applications. The results from SEM revealed that the homogeneity of the surface of the membranes relies on the amount and fusion between the two polymers. The development of microcracks in all composite samples was apparently regulated by changing the composition in the polymer ratios. FTIR verified the mixing interaction of polymers in both composites. Composite bioactivity studies showed that the highest bioactivity in these two environments existed at higher concentrations of PVA and sulfonated poly(ether ether ketone) (SPEEK). The MTT method of in vitro cytotoxicity analysis involving epithelial cells (HBL-100) revealed that excellent cell viability was depicted in all samples.

Similarly, Meer, Kausar, & Iqbal (2016) studied polymer microsphere and silica nanoparticles as effective polymer composite reinforcement additives. The focus was on their methods of production, properties, and application based on their properties. Silica is commonly used as an enhanced surface mediator, as a nucleating agent, and as cores and templates. Under the polymer–silica composites, different categories such as polypyrene, rubber, polystyrene, polyaniline, acrylate polymers, and epoxy were extensively discussed. It was found that silica nanoparticles tend to improve the mechanical properties and overall performance of polymer–silica composites. Similarly, silica-carbon nanotube composites have good mechanical and electrical properties. They are important in the application such as nanomedicines, nanoelectronic devices, and protection.

In addition, Guyard, Persello, Boisvert, & Cabane (2006) prepared a cast film composite using an aqueous solution of the polymer having silica nanoparticles. The polymer matrices used were hydroxypropyl methylcellulose (HPMC), polyvinyl alcohol (PVA), and a mixture of PVA and HPMC polymers. The polymer–silica nanoparticle interphase was investigated by adsorption isotherms in the aqueous dispersion. From the results, a high affinity for silica nanoparticles and a high adsorption coverage were observed in HPMC; in contrast, PVA had a low affinity and could adsorb at low coverage. In films, silica nanoparticle structure was observed by small-angle neutron scattering (SANS) and transmission electron microscopy (TEM). All analyses indicated that the nanoparticles of silica in HPMC films were well dispersed and aggregated in PVA films. Composite mechanical properties were assessed by tensile strength tests. In both cases, the polymers had high elastic modulus (291 MPa for PVA and 65 MPa for HPMC) and low-maximum break elongation (4.12 mm for PVA and 0.15 mm for HPMC). The inclusion of silica nanoparticles in HPMC matrix resulted in increased modulus of elasticity and decreased breaking stress. When silica nanoparticles were added in the PVA matrix, the modulus of elasticity decreased and the breaking stress increased. The polymer–silica interface modifications can be used to change the mechanical properties of these composite materials.

In a research conducted by Kierys, Dziadosz, & Goworek (2010), the monodisperse polymer–silica composite was produced by a two-stage method, with the polymer as a matrix and hydrophilic silica gel as the filler element. In the first step, Amberlite XAD7HP particle swelling was done in tetraethoxysilane (TEOS). Some amount of the TEOS-impregnated XAD7HP particles was subsequently transferred to an acidic, aqueous solution to enhance the silica precursor sol–gel process. This method is evaluated as a prospective route toward obtaining a core-shell morphology composite material. Microscopic images of scanning electron microscopy (SEM) and 29Si MAS NMR showed that silica nanofibers were formed on the polymer matrix. The silica nanoparticles were attached to the matrix of polymers. The silica shell had significantly higher mechanical properties. The polymer swelling and silica phase formation significantly altered the porosity of the original polymer material. Surprisingly, the produced composite showed much more homogenous porosity.

Additionally, Fu, Feng, Lauke, & Mai (2008) researched on the effects of the adhesion of particles, particle size, and the loading of particles on the toughness, strength, and stiffness of a variety of composites containing additives in both micro- and nano-sizes with a small aspect unit ratio. Composite toughness and strength were found to be significantly affected by all three factors, in particular particle/matrix adhesion. This could be because strength relies on an effective transmission of stress between the reinforcement additive and the matrix, noting that adhesion controls brittleness and toughness. The relationship that exists between these three factors, which in most cases coexist, has shown several trends in the influence particle loading has on the toughness and strength of the composite. The composite toughness, however, significantly varies with particle loading, not particle/matrix adhesion, because the additives have much greater modulus than the polymer matrix. The vital size of the particle, normally in nanometer, was also established, below which the stiffness of the composite is greatly improved because of the significant impact of the size of the particle, possibly caused by a "nano" effective surface area.

In a study conducted by Hussain (2018), it was reported that nanostructured particles, due to their processability, tunable properties, and low cost, are the best suited flexible materials for polymer matrix–reinforcement. Similarly, Huang, Yeh, & Lai (2012) reported that a nanocomposite containing a polymeric matrix can particularly act as an effective coating as it improves substrate surface characteristics for specific purposes. For example, a nanocomposite of polymers having an inorganic layered filler that is coated on the steel surface can greatly slow down corrosion. Besides their inherent material behavior, the simplicity and efficiency in incorporating them on substrates are the main parameters for defining effective polymer nanocomposite coatings.

3.6 CONCLUSION

Most researches have been concerned with the hydrophilic nature of polymer–silica composites. Moisture absorption is an undesirable property in most engineering applications because water causes swelling and bulking of materials, especially composites. Swelling causes dimensional changes, and it greatly affects the mechanical and thermal properties of a composite material. A lot of focus has been shifted to the nanocomposite technology with various researchers trying to lower the hydrophilicity of polymer–silica nanocomposites. A lot of research is currently being conducted to explore a wider range of naturally available silica sand or related materials to improve the hydrophilicity performance of such composites, especially for green construction materials. The currently reported research on polymer–silica composites has not been directed toward green construction application. Therefore, there is a need for researching more silica materials to reinforce polymers for green construction applications.

REFERENCES

Abdalla, F. H., Mutasher, S. A., Khalid, Y. A., Sapuan, S. M., Hamouda, A.M.S., Sahari, B. B., & Hamdan, M. M. (2007). Design and fabrication of low cost filament winding machine. *Materials & Design*, 28(1), 234–239. https://doi.org/10.1016/j.matdes.2005.06.015

Abdel-Rahman, H. A., Younes, M. M., & Yassene, A. A. M. (2018). Physico-mechanical properties of gamma-irradiated clay/polyester nanocomposites. *Polymer Composites*, 39(10), 3666–3675. https://doi.org/10.1002/pc.24395

Abliz, D., Duan, Y., Steuernagel, L., Xie, L., Li, D., & Ziegmann, G. (2013). Curing methods for advanced polymer composites-a review. *Polymers and Polymer Composites*, 21(6), 341–348.

Aghazadeh Mohandesi, J., Refahi, A., Sadeghi Meresht, E., & Berenji, S. (2011). Effect of temperature and particle weight fraction on mechanical and micromechanical properties of sand-polyethylene terephthalate composites: A laboratory and discrete element method study. *Composites Part B: Engineering*, 42(6), 1461–1467. https://doi.org/10.1016/j.compositesb.2011.04.048

Ahmed, T., & Mamat, O. (2011). The development and characterization of HDPE-silica sand nanoparticles composites. In I. Staff (Ed.), *2011 IEEE Colloquium on Humanities, Science and Engineering* (pp. 6–11). New York: IEEE. https://doi.org/10.1109/CHUSER.2011.6163824

Alagirusamy, R. (2010). Hybrid yarns for thermoplastic composites. In *Technical Textile Yarns* (pp. 387–428). Woodhead Publishing Series. Oxford: Elsevier. https://doi.org/10.1533/9781845699475.2.387

Alberto, M. (2013). Introduction of fibre-reinforced polymers–polymers and composites: Concepts, properties and processes. In M. Masuelli (Ed.), *Natural Fibre Bio-Composites Incorporating Poly (Lactic Acid)*. London: INTECH Open Access Publisher. https://doi.org/10.5772/54629

Altenbach, H., Altenbach, J., & Kissing, W. (2018). *Mechanics of Composite Structural Elements* (Second edition). Singapore: Springer.

Asmatulu, R., Khan, W. S., Reddy, R. J., & Ceylan, M. (2015). Synthesis and analysis of injection-molded nanocomposites of recycled high-density polyethylene incorporated with graphene nanoflakes. *Polymer Composites*, 36(9), 1565–1573. https://doi.org/10.1002/pc.23063

Bajuri, F., Mazlan, N., & Ishak, M. R. (2018). Water absorption analysis on impregnated kenaf with nanosilica for epoxy/kenaf composite. *IOP Conference Series: Materials Science and Engineering*, 405, 12013. https://doi.org/10.1088/1757-899X/405/1/012013

Baran, I., Cinar, K., Ersoy, N., Akkerman, R., & Hattel, J. H. (2017). A review on the mechanical modeling of composite manufacturing processes. *Archives of Computational Methods in Engineering*, 24(2), 365–395. https://doi.org/10.1007/s11831-016-9167-2

Casati, R., & Vedani, M. (2014). Metal matrix composites reinforced by nano-particles—A review. *Metals*, 4(1), 65–83. https://doi.org/10.3390/met4010065

Choi, S.-J., Lee, S.-H., Ha, Y.-C., Yu, J.-H., Doh, C.-H., Lee, Y., … Shin, H.-C. (2018). Synthesis and electrochemical characterization of a glass-ceramic Li 7 P 2 S 8 I solid electrolyte for all-solid-state Li-ion batteries. *Journal of the Electrochemical Society*, 165(5), A957–A962. https://doi.org/10.1149/2.0981805jes

Chowdhury, T. U., Amin, M. M., Haque, K. A., & Rahman, M. M. (2018). A review on the use of polyethylene terephthalate (PET) as aggregates in concrete. *Malaysian Journal of Science*, 37(2), 118–136. https://doi.org/10.22452/mjs.vol37no2.4

Dawoud, M. M., & Saleh, H. M. (2019). Introductory chapter: Background on composite materials. In H. E.-D. M. Saleh & M. Koller (Eds.), *Characterizations of Some Composite Materials*. London: IntechOpen. https://doi.org/10.5772/intechopen.80960

Drakonakis, V. M., Seferis, J. C., & Doumanidis, C. C. (2013). Curing pressure influence of out-of-autoclave processing on structural composites for commercial aviation. *Advances in Materials Science and Engineering, 2013*(7), 1–14. https://doi.org/10.1155/2013/356824

Finkbeiner, M. (2013). From the 40s to the 70s—The future of LCA in the ISO 14000 family. *The International Journal of Life Cycle Assessment, 18*(1), 1–4. https://doi.org/10.1007/s11367-012-0492-x

Friedrich, K., & Almajid, A. A. (2013). Manufacturing aspects of advanced polymer composites for automotive applications. *Applied Composite Materials, 20*(2), 107–128. https://doi.org/10.1007/s10443-012-9258-7

Fu, S.-Y., Feng, X.-Q., Lauke, B., & Mai, Y.-W. (2008). Effects of particle size, particle/matrix interface adhesion and particle loading on mechanical properties of particulate–polymer composites. *Composites Part B: Engineering, 39*(6), 933–961. https://doi.org/10.1016/j.compositesb.2008.01.002

Gand, A. K., Chan, T.-M., & Mottram, J. T. (2013). Civil and structural engineering applications, recent trends, research and developments on pultruded fiber reinforced polymer closed sections: A review. *Frontiers of Structural and Civil Engineering, 7*(3), 227–244. https://doi.org/10.1007/s11709-013-0216-8

Ghori, S. W., Siakeng, R., Rasheed, M., Saba, N., & Jawaid, M. (2018). The role of advanced polymer materials in aerospace. In *Sustainable Composites for Aerospace Applications* (pp. 19–34). Woodhead Publishing Series in Composites Science and Engineering. Oxford: Elsevier. https://doi.org/10.1016/B978-0-08-102131-6.00002-5

Guyard, A., Persello, J., Boisvert, J.-P., & Cabane, B. (2006). Relationship between the polymer/silica interaction and properties of silica composite materials. *Journal of Polymer Science Part B: Polymer Physics, 44*(7), 1134–1146. https://doi.org/10.1002/polb.20768

Hamidi, Y. K., & Altan, M. C. (2017). Process induced defects in liquid molding processes of composites. *International Polymer Processing, 32*(5), 527–544. https://doi.org/10.3139/217.3444

Hernández, S., Sket, F., Molina-Aldareguı́a, J. M., González, C., & LLorca, J. (2011). Effect of curing cycle on void distribution and interlaminar shear strength in polymer-matrix composites. *Composites Science and Technology, 71*(10), 1331–1341. https://doi.org/10.1016/j.compscitech.2011.05.002

Herrera-Franco, P., Valadez-Gonzalez, A., & Cervantes-Uc, M. (1997). Development and characterization of a HDPE-sand-natural fiber composite. *Composites Part B: Engineering, 28*(3), 331–343. https://doi.org/10.1016/S1359-8368(96)00024-8

Holbery, J., & Houston, D. (2006). Natural-fiber-reinforced polymer composites in automotive applications. *JOM, 58*(11), 80–86. https://doi.org/10.1007/s11837-006-0234-2

Huang, T.-C., Yeh, J.-M., & Lai, C.-Y. (2012). Polymer nanocomposite coatings. In *Advances in Polymer Nanocomposites* (pp. 605–638). Woodhead Publishing Limited. Oxford: Elsevier. https://doi.org/10.1533/9780857096241.3.605

Hussain, C. M. (2018). Polymer nanocomposites—An intro. In *New Polymer Nanocomposites for Environmental Remediation* (pp. xxi–xxv). Woodhead Publishing Series. Oxford: Elsevier. https://doi.org/10.1016/B978-0-12-811033-1.09997-2

Irving, P. E., & Soutis, C. (2015). *Polymer Composites in the Aerospace Industry*. Woodhead Publishing series in composites science and engineering: Number 50. Cambridge: Woodhead Publishing.

Ji, X., Hampsey, J. E., Hu, Q., He, J., Yang, Z., & Lu, Y. (2003). Mesoporous silica-reinforced polymer nanocomposites. *Chemistry of Materials, 15*(19), 3656–3662. https://doi.org/10.1021/cm0300866

Kalambettu, A., Venkatesan, P., & Dharmalingam, S. (2012). Polymer/silica composites fabricated by sol-gel technique for medical applications. *Trends in Biomaterials and Artificial Organs: An International Journal, 26*(3), 121–129.

Katoh, M., & Nakagama, H. (2014). FGF receptors: Cancer biology and therapeutics. *Medicinal Research Reviews, 34*(2), 280–300. https://doi.org/10.1002/med.21288

Khan, L. A., Khan, W. A., & Ahmed, S. (2017). Out-of-autoclave (OOA) manufacturing technologies for composite sandwich structures. In R. Das & M. Pradhan (Eds.), *Handbook of Research on Manufacturing Process Modeling and Optimization Strategies* (Vol. 34, pp. 292–317). Hershey, PA: IGI Global. https://doi.org/10.4018/978-1-5225-2440-3.ch014

Kierys, A., Dziadosz, M., & Goworek, J. (2010). Polymer/silica composite of core-shell type by polymer swelling in TEOS. *Journal of Colloid and Interface Science, 349*(1), 361–365. https://doi.org/10.1016/j.jcis.2010.05.049

Koniuszewska, A. G., & Kaczmar, J. W. (2016). Application of polymer based composite materials in transportation. *Progress in Rubber Plastics and Recycling Technology, 32*(1), 1–24. https://doi.org/10.1177/147776061603200101

Krasucka, P., Stefaniak, W., Kierys, A., & Goworek, J. (2015). Polymer–silica composites and silicas produced by high-temperature degradation of organic component. *Thermochimica Acta, 615*, 43–50. https://doi.org/10.1016/j.tca.2015.07.004

Mack, J., & Schledjewski, R. (2012). Filament winding process in thermoplastics. In *Manufacturing Techniques for Polymer Matrix Composites (PMCs)* (pp. 182–208). Woodhead Publishing Series in Composites Science and Engineering. Oxford: Elsevier. https://doi.org/10.1533/9780857096258.2.182

Macke, A., Schultz, B., & Rohatgi, P. K. (2012). Metal matrix composites offer the automotive industry an opportunity to reduce vehicle weight, improve performance. *Advanced Materials and Processes, 170*, 19–23.

Madhav, H., Singh, N., & Jaiswar, G. (2019). Thermoset, bioactive, metal–polymer composites for medical applications. In A. M. Grumezescu & V. Grumezescu (Eds.), *Materials for Biomedical Engineering. Thermoset and Thermoplastic Polymers* (pp. 105–143). Amsterdam: Elsevier. https://doi.org/10.1016/B978-0-12-816874-5.00004-9

Maria, M. (2013). Advanced composite materials of the future in aerospace industry. *INCAS Bulletin, 5*(3), 139–150. https://doi.org/10.13111/2066-8201.2013.5.3.14

Medina, N. F., Garcia, R., Hajirasouliha, I., Pilakoutas, K., Guadagnini, M., & Raffoul, S. (2018). Composites with recycled rubber aggregates: Properties and opportunities in construction. *Construction and Building Materials, 188*, 884–897. https://doi.org/10.1016/j.conbuildmat.2018.08.069

Meer, S., Kausar, A., & Iqbal, T. (2016). Attributes of polymer and silica nanoparticle composites: A review. *Polymer-Plastics Technology and Engineering, 55*(8), 826–861. https://doi.org/10.1080/03602559.2015.1103267

Minsch, N., Herrmann, F. H., Gereke, T., Nocke, A., & Cherif, C. (2017). Analysis of filament winding processes and potential equipment technologies. *Procedia CIRP, 66*, 125–130. https://doi.org/10.1016/j.procir.2017.03.284

Mitschang, P., & Hildebrandt, K. (2012). Polymer and composite moulding technologies for automotive applications. In *Advanced Materials in Automotive Engineering* (pp. 210–229). Woodhead Publishing Series in Composites Science and Engineering. Oxford: Elsevier. https://doi.org/10.1533/9780857095466.210

Morris, H. (2018). Fat taxes and skinny flight attendants: The crazy ways airlines lose weight. Retrieved from https://www.traveller.com.au/airline-weight-reduction-to-save-fuel-the-crazy-ways-airlines-save-weight-on-planes-h14vlh

Mosallam, A. S. (2014). Polymer composites in construction: An overview. *SOJ Materials Science & Engineering, 2*(1), 1–25. https://doi.org/10.15226/sojmse.2014.00107

Murr, L. E. (2014). *Handbook of Materials Structures, Properties, Processing and Performance.* New York: Springer.

Neşer, G. (2017). Polymer based composites in marine use: History and future trends. *Procedia Engineering, 194*, 19–24. https://doi.org/10.1016/j.proeng.2017.08.111

Nielsen, R. (2005). Molecular signatures of natural selection. *Annual Review of Genetics, 39*, 197–218. https://doi.org/10.1146/annurev.genet.39.073003.112420

Ogale, A., & Schlimbach, J. (2011). Quickstep: Beyond Out of Autoclave Curing. *JEC Composites, 36*, 62–64

Ouarhim, W., Zari, N., Bouhfid, R., & Qaiss, A. E. K. (2018). Mechanical performance of natural fibers–based thermosetting composites. In M. Jawaid, M. Thariq, & N. Saba (Eds.), *Mechanical and Physical Testing of Biocomposites, Fibre-Reinforced Composites and Hybrid Composites*. Woodhead Publishing series in composites science and engineering (pp. 43–60). Oxford: Woodhead Publishing. https://doi.org/10.1016/B978-0-08-102292-4.00003-5

Peng, Z., & Kong, L. X. (2007). Morphology of self-assembled polyvinyl alcohol/silica nanocomposites studied with atomic force microscopy. *Polymer Bulletin, 59*(2), 207–216. https://doi.org/10.1007/s00289-007-0756-y

Pickering, K. L., Efendy, M. A., & Le, T. M. (2016). A review of recent developments in natural fibre composites and their mechanical performance. *Composites Part A: Applied Science and Manufacturing, 83*, 98–112. https://doi.org/10.1016/j.compositesa.2015.08.038

Porwal, H., Tatarko, P., Grasso, S., Khaliq, J., Dlouhý, I., & Reece, M. J. (2013). Graphene reinforced alumina nano-composites. *Carbon, 64*, 359–369. https://doi.org/10.1016/j.carbon.2013.07.086

Qi, B., Yuan, Z., Lu, S., Liu, K., Li, S., Yang, L., & Yu, J. (2014). Mechanical and thermal properties of epoxy composites containing graphene oxide and liquid crystalline epoxy. *Fibers and Polymers, 15*(2), 326–333. https://doi.org/10.1007/s12221-014-0326-5

Rahaman, M., Khastgir, D., & Aldalbahi, A. K. (2019). *Carbon-Containing Polymer Composites*. Springer series on polymer and composite materials. Cham, Switzerland: Springer.

Rahman, M. R., Chang Hui, J. L., & Hamdan, S. B. (2018). Introduction and reinforcing potential of silica and various clay dispersed nanocomposites. In M. R. Rahman (Ed.), *Silica and Clay Dispersed Polymer Nanocomposites: Preparation, Properties and Applications*. Woodhead Publishing series in composites science and engineering. (pp. 1–24). Oxford: Woodhead Publishing. https://doi.org/10.1016/B978-0-08-102129-3.00001-4

Ramakrishna, S., Mayer, J., Wintermantel, E., & Leong, K. W. (2001). Biomedical applications of polymer-composite materials: A review. *Composites Science and Technology, 61*(9), 1189–1224. https://doi.org/10.1016/S0266-3538(00)00241-4

Ramnath, B. V., Elanchezhian, C., Annamalai, R. M., Aravind, S., Atreya, T. S. A., Vignesh, V., & Subramanian, C. (2014). Aluminium metal matrix composites—A review. *Reviews on Advanced Materials Science, 38*(5), 55–60.

Reichenbach, C., Kalies, G., Enke, D., & Klank, D. (2011). Cavitation and pore blocking in nanoporous glasses. *Langmuir: The ACS Journal of Surfaces and Colloids, 27*(17), 10699–10704. https://doi.org/10.1021/la201948c

Seghiri, M., Boutoutaou, D., Kriker, A., & Hachani, M. I. (2017). The possibility of making a composite material from waste plastic. *Energy Procedia, 119*, 163–169. https://doi.org/10.1016/j.egypro.2017.07.065

Singh, D., Chauhan, N. P. S., Mozafari, M., & Hiran, B. L. (2016). High-temperature resistive free radically synthesized chloro-substituted phenyl maleimide antimicrobial polymers. *Polymer-Plastics Technology and Engineering, 55*(18), 1916–1939. https://doi.org/10.1080/03602559.2016.1185620

Slieptsova, I., Savchenko, B., Sova, N., & Slieptsov, A. (2016). Polymer sand composites based on the mixed and heavily contaminated thermoplastic waste. *IOP Conference Series: Materials Science and Engineering, 111*, 12027. https://doi.org/10.1088/1757-899X/111/1/012027

Sreeprasad, T. S., Maliyekkal, S. M., Lisha, K. P., & Pradeep, T. (2011). Reduced graphene oxide-metal/metal oxide composites: Facile synthesis and application in water purification. *Journal of Hazardous Materials, 186*(1), 921–931. https://doi.org/10.1016/j.jhazmat.2010.11.100

Sultana, R., Akter, R., Alam, M. Z., Qadir, M. R., Begum, M. A., & Gafur, M. A. (2013). Preparation and characterization of sand reinforced polyester composites. *International Journal of Engineering & Technology IJET-IJENS*, *13*(2), 111–118.

Thommes, M. (2010). Physical adsorption characterization of nanoporous materials. *Chemie Ingenieur Technik*, *82*(7), 1059–1073. https://doi.org/10.1002/cite.201000064

Walbran, W. A., Bickerton, S., & Kelly, P. A. (2013). Evaluating the shear component of reinforcement compaction stress during liquid composite moulding processes. *Journal of Composite Materials*, *47*(5), 513–528. https://doi.org/10.1177/0021998312441811

Walker, L. S., Marotto, V. R., Rafiee, M. A., Koratkar, N., & Corral, E. L. (2011). Toughening in graphene ceramic composites. *ACS Nano*, *5*(4), 3182–3190. https://doi.org/10.1021/nn200319d

Wang, X.-B., Jovel, J., Udomporn, P., Wang, Y., Wu, Q., Li, W.-X., ... Ding, S.-W. (2011). The 21-nucleotide, but not 22-nucleotide, viral secondary small interfering RNAs direct potent antiviral defense by two cooperative argonautes in *Arabidopsis thaliana*. *The Plant Cell*, *23*(4), 1625–1638. https://doi.org/10.1105/tpc.110.082305

Wei, R., Hua, X., & Xiong, Z. (2018). Polymers and polymeric composites with electronic applications. *International Journal of Polymer Science*, *2018*, 1. https://doi.org/10.1155/2018/8412480

Witik, R. A., Payet, J., Michaud, V., Ludwig, C., & Månson, J.-A. E. (2011). Assessing the life cycle costs and environmental performance of lightweight materials in automobile applications. *Composites Part A: Applied Science and Manufacturing*, *42*(11), 1694–1709. https://doi.org/10.1016/j.compositesa.2011.07.024

Younes, M. M., El-sharkawy, I. I., Kabeel, A. E., Uddin, K., Pal, A., Mitra, S., ... Saha, B. B. (2019). Synthesis and characterization of silica gel composite with polymer binders for adsorption cooling applications. *International Journal of Refrigeration*, *98*, 161–170. https://doi.org/10.1016/j.ijrefrig.2018.09.003

Zafar, M., Najeeb, S., Khurshid, Z., Vazirzadeh, M., Zohaib, S., Najeeb, B., & Sefat, F. (2016). *Potential of electrospun nanofibers for biomedical and dental applications. Materials (Basel, Switzerland)*, *9*(2), 73. https://doi.org/10.3390/ma9020073

Zagho, M. M., Hussein, E. A., & Elzatahry, A. A. (2018). Recent overviews in functional polymer composites for biomedical applications. *Polymers*, *10*(7), 739. https://doi.org/10.3390/polym10070739

Zhang, C., Chen, Y., Li, H., Tian, R., & Liu, H. (2018). Facile fabrication of three-dimensional lightweight RGO/PPy nanotube/Fe_3O_4 aerogel with excellent electromagnetic wave absorption properties. *ACS Omega*, *3*(5), 5735–5743. https://doi.org/10.1021/acsomega.8b00414

4 The Role and Applications of Nanomaterials in the Automotive Industry

V. Dhinakaran and M. Varsha Shree
Chennai Institute of Technology

CONTENTS

4.1	Introduction	51
4.2	Carbon Nanotubes for Automobile Radiators	52
4.3	CNTs for Pressure Sensing	53
4.4	Silicon Carbide	54
4.5	Si-Al for Automobiles	55
4.6	Graphene	55
4.7	Nano Coatings	56
4.8	Scratch-Resistant Nanocoatings	57
4.9	Nano Varnish	57
4.10	Conclusions	58
References		58

4.1 INTRODUCTION

In transportation, nanotechnology offers a variety of advantages, including improving the strength and durability of cars over a longer period. To enhance their performance and sustainability, nanotechnology may be used for different body parts, including chassis, tires, windows, and motors. Nanotechnology's potential applications in transport vehicles are almost endless. Nanomaterials, nanostructures, and nanodevices are being developed and manufactured as innovative ways to build robust vehicles. Nanotechnology is used to protect vehicle bodies from corrosion and to provide abrasion resistance as a practical tool. Nanotechnology is the atomic and molecular engineering of matter [1]. It generally works with materials, devices, and other structures of size ranging from 1 to 100nm. Scientists and engineers now consider a wide variety of ways to manufacture substances in nanogroups and exploit their improved properties, such as high intensity, light weight, increased light spectrum exposure, and increased chemical reactivity [2]. The overall production of their software, engineering, logistics, and technology are the main aspects. Materials with nanoscopic dimensions are

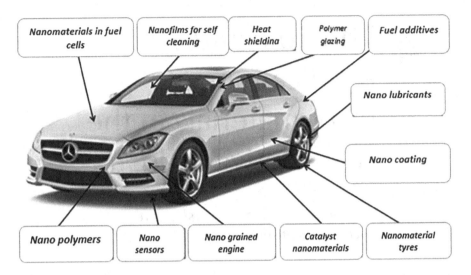

FIGURE 4.1 Nanotechnology in automobiles [4].

of fundamental interest, not only in the area of device technology and drug delivery but in the process of transition between the bulk and the molecular scale, to have potential technological applications. Superior biological, electrical, electronic, magnetic, and quantum mechanical properties are exhibited by nanomaterials. As such, a number of nanoscale materials, including nanoparticles, nanofibers, nanocomposites, nanofilms, and other products, have been used in many different industries because of these superior properties [3]. These materials are considered to be the next generation of nanoscale materials. In over 1100 different products, including concrete, automobiles, polymer coating, tennis rackets, wrinkle-resistant clothes, and other optical, electronic, diagnostic, and sensing systems, nanomaterials have already been found in many different fields as represented in Figure 4.1.

4.2 CARBON NANOTUBES FOR AUTOMOBILE RADIATORS

The trend toward higher motor capacity leads to larger car radiators and higher frontal areas, resulting in higher fuel consumption. To optimize fuel consumption, heat transfer of the coolant flow via automotive radiators is of great importance.

Compaq can be created by injecting nano-coolants into the car radiator. The convective heat transfer enhancement of carbon nanotube (CNT)–water nanofluid has been studied experimentally inside an automobile radiator [5]. To optimize fuel consumption, the heat recovery efficiency of the coolant through vehicle radiators is highly significant. Four different nanofluid concentrations in the range of 0.15–1 volume in this study. The addition of CNT nanoparticles to water has prepared a number. Conventional liquids are found in high-energy applications including automotive motors to meet the growing need for cooling. CNTs have a higher thermal conductivity, a more elevated appearance ratio, a lower special gravity, a greater specific surface area (SSA), and lower thermal resistance compared to Al_2O_3–water and CuO–water nanofluids [6]. The use of Al_2O_3–water nanofluid as a jacket coolant in diesel electrical

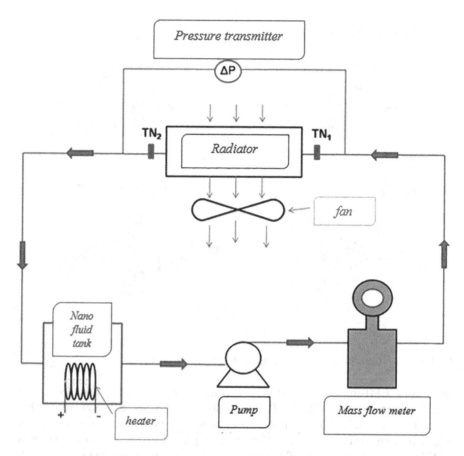

FIGURE 4.2 Nanofluid radiator [7].

generators improved the performance of cogeneration. The decline in real heat is due to a drop in the removal of the surplus heat from the generator. The construction of nanofluid radiators is depicted in Figure 4.2. But due to their superior convective thermal transfer coefficient, waste heat recovery with the use of nanofluids improved [8].

4.3 CNTs FOR PRESSURE SENSING

The so-called black pollution (WTR) is a strong environmental influence because of the immense use of tires. As the need for more and diverse sensors is growing, printed electronics have been described as a promising solution for inexpensive, omnipresent sensor systems. This study demonstrates how CNT-based thin-film transistors (CNT-TFTs) are used to sense environmental pressure from 0 to 42 psig over a pressurized range.

The CNT-TFT transductance is linearly linked to environmental pressure with a sensitivity of 48.1 pS/psi. The rapid development of flexible electronics such as wearable health surveillance systems, flexible displays, or flexible power supply systems means

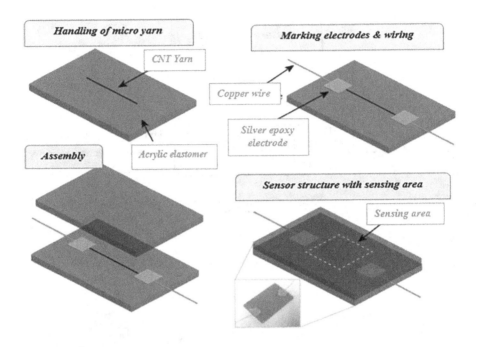

FIGURE 4.3 Construction of carbon nanotube pressure sensor [10].

that flexible EMI-shielding materials have a large potential for the market [9]. In this regard, the ground tire rubber (GTR), using its three-dimensional (3D) cross-linked structure, was completely converted into a valuable, high-performance electromagnetic interference (EMI) blind material. There was a typical different setup of CNTs solely on GTR borders of the CNT/GTR. The sensor infrastructure required to facilitate the rapid expansion of the Internet of Things (IoT) has become a key approach to low cost, omnipresent sensor networks. Thin-film carbon nanotubes (CNT-TFTs) for the sensing of environmental pressures across a wide pressure spectrum for an advanced tire sensor device that is completely imprinted capable of measuring differential pneumatic pressure and depth is represented in Figure 4.3 [11].

4.4 SILICON CARBIDE

Composites strengthened by an aluminum matrix with SiC particles have improved friction and wear compared to standard cast iron as well as have a higher heat power and heat conductivity and in particular less bulk. Some of the important uses of metal matrix composites (MMCs) is as the brake rotor containing Al-SiC particle composites in automotive brake systems. Two Al-Si alloys containing 10%–20% SiC dry particles reinforced with a semi-metal traction material with wear behavior. It was found that the friction coefficient is very high for charges less than 200N by abrasion and adhesion (around 0.45). For loads over 200N, the coefficient of friction decreases when the load in the two materials increases, while the wear rate increases with an increase in the load. SiC devices are expected to result in much greater energy savings in sophisticated power electronics applications. High current power (>100A) devices and high-temperature operations are typically required for automotive applications [12]. SiC modules using the nickel micro plating bonding (NMPB) technique were tested for their viability as a power module in automotive applications, where the narrow gap between the SiC unit and the substratum can be securely bound by Ni plating. NMPB's TO247 SiC-SBD module gives a greater bonding strength than standard lead-free solder, even after high-temperature storage, for 1,000h and after 1,000 cycles of temperature cycle testing. The bonding strength is 250°C.

4.5 Si-Al FOR AUTOMOBILES

The growing demand in automotive industries for weight reduction, energy savings, and pollution reduction has led to the development of new lightweight automobile materials. Mg_2Si composites are a good candidate as a vehicle split disk substance because the intermetallic combination Mg_2Si exhibits high fusing, low density, moderate stiffness, low thermo-expansion coefficient, and relatively high elastic modulus. Hypereutectic Al-Si alloys are an in situ aluminum matrix that contains a significant number of hard particles of Mg_2Si. In an automotive exhaust system, catalytic converters are mounted. After the mid-1970s, they have always been more used to meet pollution restrictions laid down by international laws. Platinum group metals (PGM) are used as a buffer, reducing undesired exhaust gas (CO) as the major by-products of internal combustion processes, such as unburned carbohydrate (HC) and nitrogen oxides (NOx). As for the automotive brake disk material, the Mg_2Si/Al-Si-Cu composite offers a high melting temperature of 1085 BC low density of 103 kg per Mg_2Si–Cu component. Many Mg_2Si composites are generally very coarse in normal as-cast Mg_2Si/Al-Si, resulting in poor properties [13]. Therefore, they must be modified to ensure adequate mechanical strength and ductility with the mainly ground primary Mg_2Si crystals.

4.6 GRAPHENE

Friction and corrosion of damaged materials are a significant source of energy dissipation in vehicle motors. Owing to their tribological behavior, graphene (Gr) nanolubricants save energy and reduce exhaust emissions. Tribometers based on ASTMG181 are tested for the antifriction and wear properties of Gr nanolubricants. In this context, we discuss the self-healing process of tribological events. Gr nanolubricants improve antifriction and antiwear by 29%–35% and 22%–29%, respectively, in the border lubrication system. During NEDC testing, a 17% reduction in cumulative fuel consumption was observed by lubricating the motor using Gr nanolubricants. Lubricants are used primarily to eliminate sliding movement between two parts, which is represented in Figure 4.4 [15]. Lubricants are mainly used in internal-combustion engines, automobiles and automotive gearboxes, compressors, generators, and hydraulic systems, and account for about 50% of the global demand for motor-oil applications. Recently, metallic nanoparticles added to lubricant oils (nanolubricants) have been reported to act as wear-resistant additives under extreme pressures. Metal nanoparticles are not corrosive and can be used at very high temperatures. So the beginning of a new age of antiwear and extreme pressure additives

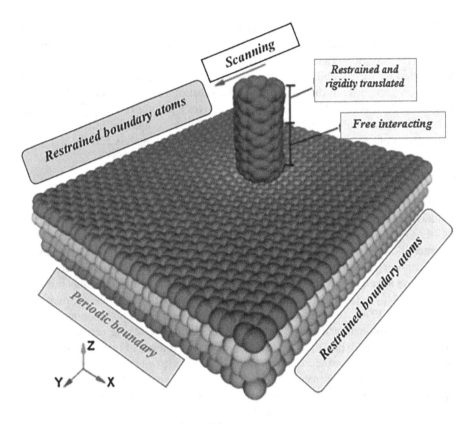

FIGURE 4.4 Graphene nanolubricant [14].

is very exciting. The extreme thermal stability of Al_2O_3/TiO_2 nanolubricants enables the development of different types of heat transfer nanofluids with high efficiency, improving lubricating oil drain cycles and reducing repair costs in vehicles in the wide spectrum of efficient applications [16].

4.7 NANO COATINGS

Coating today is not only meant to enhance efficiency but also serves as a safeguard against the deterioration of precious metals and structures, which account for almost 4% of GNP worldwide. Nanocoatings are materials generated to create a more compact component by reducing the contents at the molecular level. The appearance and usability of nanoparticles provide the painting and covering industry a number of advantages and opportunities. Nanocoating, including chemical vapor deposition, solid vapor deposition, is possible in a variety of ways. The accumulation of electrochemical sol-gel methods, electrical exposure, and surface laser beam treatment. A new type of paint has been developed and characterized by the dispersion of multi-walled carbon nanotubes (MWCNTs) using a polymer matrix. CNTs have outstanding electrical, thermal, optical, and mechanical properties [17]. The fresh colors, which spread small percentages (from 0.5 to 3wt.%) in epoxy-resin based paints, have been developed and characterized by electrically conductive characteristics. The nanofiller dispersion process in a polymer matrix was performed using the technique of three roller mills, known as high shear forces, with satisfactory results. The magnitude values obtained were up to 10^{-2} S/cm depending on the MWCNT content.

4.8 SCRATCH-RESISTANT NANOCOATINGS

The right combination of hardness and flexibility is required to achieve optimal scratch resistance. In this context, the development of scratch-resistant covering is achieved by organic and inorganic films as represented in Figure 4.5. Recent developments in the production of scratch-resistant coatings in nanotechnology have been significant. To improve scratch- and abrasion-prone coverings, Glasel et al. demonstrated the use of SiO_2 siloxane-encapsulated nanoparticles. By incorporating SiO_2 nanoparticles in an organic matrix that can pass to the surface, the coating industries have developed scrabble-resistant coatings. This increases the scratch resistance by enriching nanoparticles close to the cover surface. These nano-synthesized ceramics are used to improve corrosion, wear, and scratch protection of chemically active and mechanically soft materials with superior chemical inertness and hardness in the area of electroless nickel coating. Nevertheless, coatings are mainly determined by the surface state and microstructures and their electrochemical and mechanical properties [19]. The nanotechnology that received great interest is nanocoating for motor applications in the automotive industry.

4.9 NANO VARNISH

A pressing problem for a denture is that during its working and cleaning (e.g., using a brush), it undergoes abrasive wear. As a result, the surface hardness of the acrylic coating is weakened over time, and the dental base materials are exposed to both mechanical and chemical applications. It is found that the highest surface hardness and elastic modulus are obtained by silica-nano products that contain surfactants, but no statistical significance was observed during aging for 6–12 months [20]. Despite improvements

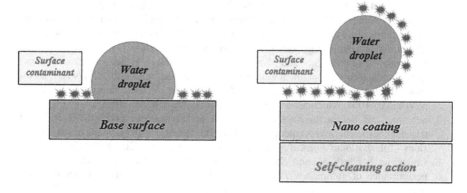

FIGURE 4.5 Comparison of nanocoated surfaces [18].

observed by scanning electron microscopy, the surface ruggedness has not improved substantially over time for any group, contrary to standards. Increased resin surface strength is reported for silica nanoparticles (SiO_2), while their low surface energy characteristic makes adherence difficult for biofilms. For acrylic resin specimens coated with OPTIGLAZE and nano varnish respectively for contrast with other grades, statistically higher hardness and elastic modulus values were obtained [21].

4.10 CONCLUSIONS

Designed for automotive applications, nanomaterials are used particularly for safety and reducing energy consumption. The reduction in weight of the cars is a fair outcome of nanomaterial use and the structures being built from low-density materials. Nanotechnology can make use of materials and resources more efficiently, thus reducing waste and emissions. In the automotive industry, various forms of nanomaterials, such as nanoparticles, nanofilms, nanoplasms, nanofibers, and nanocomposites, have been used to enhance the mechanical, electronic, thermal, corrosive, self-cleaning, and antiwear properties and sensing abilities. While these materials exhibit superior characteristics for different automotive applications, they can be dangerous if not properly handled. They can be manufactured at a low cost and can have certain performance and process benefits in comparison to metals and polymer composites commonly used in the manufacturing of parts for automobile applications.

REFERENCES

1. Jadar, R., K. S. Shashishekar, and S. R. Manohara. "Nanotechnology integrated automobile radiator." *Materials Today: Proceedings* 4, no. 11 (2017): 12080–12084.
2. Sequeira, S. "Applications of nanotechnology in automobile industry." (2015). DOI: 10.13140/RG.2.1.3821.8960.
3. Stella, A. Josephine, and K. Rajeswari. "Impact of nanotechnology in automobile industry." *ZENITH International Journal of Business Economics & Management Research* 2, no. 12 (2012): 298–302.
4. Asmatulu, Ramazan, P. Nguyen, and Eylem Asmatulu. "Nanotechnology safety in the automotive industry." In *Nanotechnology Safety*, pp. 57–72. Elsevier, 2013. ISBN 9780444594389, http://dx.doi.org/10.1016/B978-0-444-59438-9.00005-9.
5. Chougule, Sandesh S., and Santosh Kumar Sahu. "Thermal performance of automobile radiator using carbon nanotube-water nanofluid—experimental study." *Journal of Thermal Science and Engineering Applications* 6, no. 4 (2014): 1–6.
6. Chougule, Sandesh S., and Santosh Kumar Sahu. "Experimental investigation of heat transfer augmentation in automobile radiator with CNT/water nanofluid." In *International Conference on Micro/Nanoscale Heat Transfer*, vol. 36154, p. V001T02A008. American Society of Mechanical Engineers, 2013.
7. Selvam, C., R. Solaimalai Raja, D. Mohan Lal, and Sivasankaran Harish. "Overall heat transfer coefficient improvement of an automobile radiator with graphene based suspensions." *International Journal of Heat and Mass Transfer* 115 (2017): 580–588.
8. Chougule, Sandesh S., and Santosh Kumar Sahu. "Comparative study of cooling performance of automobile radiator using Al_2O_3-water and carbon nanotube-water nanofluid." *Journal of Nanotechnology in Engineering and Medicine* 5, no. 1 (2014): 5. Doi: 011001-1-011001-5.

9. Andrews, Joseph B., Jorge A. Cardenas, Chin Jie Lim, Steven G. Noyce, Jacob Mullett, and Aaron D. Franklin. "Fully printed and flexible carbon nanotube transistors for pressure sensing in automobile tires." *IEEE Sensors Journal* 18, no. 19 (2018): 7875–7880.
10. Dinh, Toan, Tuan-Khoa Nguyen, Hoang-Phuong Phan, Jarred Fastier-Wooller, Canh-Dung Tran, Nam-Trung Nguyen, and Dzung Viet Dao. "Fabrication of a sensitive pressure sensor using carbon nanotube micro-yarns." In *2017 IEEE SENSORS*, pp. 1–3. IEEE, 2017.
11. Jing, Weijie, Chao Yang, Yin Wu, Qiang Zhao, Li Chen, and Gang Li. "CNT-coated magnetic self-assembled elastomer micropillar arrays for sensing broad-range pressures." *Nanotechnology* 31, no. 43 (2020): 435501.
12. Qian, Mengbo, Xiaodong Xu, Zhe Qin, and Shaoze Yan. "Silicon carbide whiskers enhance mechanical and anti-wear properties of PA6 towards potential applications in aerospace and automobile fields." *Composites Part B: Engineering* 175 (2019): 107096.
13. Liu, Xiaobo, Miao Yang, Dekun Zhou, and Yuguang Zhao. "Microstructure and wear resistance of Mg_2Si–Al composites fabricated using semi-solid extrusion." *Metals* 10, no. 5 (2020): 596.
14. Berman, Diana, Ali Erdemir, and Anirudha V. Sumant. "Graphene: a new emerging lubricant." *Materials Today* 17, no. 1 (2014): 31–42.
15. Izzaty, N., and H. Y. Sastra. "The implementation of graphene composites for automotive: an industrial perspective." In *IOP Conference Series: Materials Science and Engineering*, vol. 536, no. 1, p. 012133. IOP Publishing, 2019.
16. Nomura, Keita, Hirotomo Nishihara, Naoya Kobayashi, Toshihiro Asada, and Takashi Kyotani. "4.4 V supercapacitors based on super-stable mesoporous carbon sheet made of edge-free graphene walls." *Energy & Environmental Science* 12, no. 5 (2019): 1542–1549.
17. Behera, Ajit, P. Mallick, and S. S. Mohapatra. "Nanocoatings for anticorrosion: An introduction." In *Corrosion Protection at the Nanoscale*, pp. 227–243. Elsevier, 2020.
18. https://www.azonano.com/image.axd?src=%2fimages%2fArticle_Images%2fImageForArticle_3031(1).jpg&ts=20120622114846&ri=499.
19. Li, Yongqiang, Ling Zhang, and Chunzhong Li. "Highly transparent and scratch resistant polysiloxane coatings containing silica nanoparticles." *Journal of Colloid and Interface Science* 559 (2020): 273–281.
20. Tirupathi, Sunnypriyatham, S. V. S. G. Nirmala, Srinitya Rajasekhar, and Sivakumar Nuvvula. "Comparative cariostatic efficacy of a novel nano-silver fluoride varnish with 38% silver diamine fluoride varnish a double-blind randomized clinical trial." *Journal of Clinical and Experimental Dentistry* 11, no. 2 (2019): e105.
21. Nozari, Ali, Shabnam Ajami, Azade Rafiei, and Elmira Niazi. "Impact of nano hydroxyapatite, nano silver fluoride and sodium fluoride varnish on primary teeth enamel remineralization: an in vitro study." *Journal of Clinical and Diagnostic Research: JCDR* 11, no. 9 (2017): ZC97.

5 Characterization Tools and Techniques for Nanomaterials and Nanocomposites

Ruma Arora Soni
Maulana Azad National Institute of Technology

R. S. Rana
Maulana Azad National Institute of Technology

S. S. Godara
RTU

CONTENTS

5.1	Introduction	62
5.2	Characterization Techniques	63
	5.2.1 Diffraction	63
	5.2.1.1 X-Ray Diffraction	63
	5.2.1.2 Neutron Diffraction	65
	5.2.2 Microscopy	68
	5.2.2.1 Electron Microscopy	68
	5.2.3 Spectroscopy	70
	5.2.3.1 FTIR Spectroscopy	70
	5.2.3.2 UV–Visible Absorption Spectroscopy	71
	5.2.3.3 Raman Spectroscopy	72
	5.2.3.4 Nuclear Magnetic Resonance	74
	5.2.4 Thermal Mechanical Analysis (TMA)	76
	5.2.4.1 Differential Scanning Calorimetry (DSC)	76
	5.2.4.2 Thermogravimetric Analysis (TGA)	78
	5.2.4.3 Dynamic Mechanical Thermal Analysis (DMTA)	78
5.3	Conclusion	80
References		80

5.1 INTRODUCTION

The combination of two distinct materials, e.g., polymeric, is an easy way to integrate the desirable aspects of the various materials to improve the poor quality of a single material. In the world around us, there are many different examples of composite materials. For example, wood and bone are natural compounds [1–3]. The addition of hyperplatted polymers and inorganic nanofillers, for example, in a polymeric mix of organic and inorganic fillers is a new and promising example of improved properties that can be achieved by different processes. The researchers' interest was especially because of the unexpected synergistic effects derived from either component, which were enhanced by nanometer and inorganic dimensional fillets called nanocomposites. The most widely studied nanomalytic polymers, PNs, and modified boehmite or carbon nanotubes (CNTs) are thermoplastic or thermosetting matrices. Polymer/clay nanocomposites are characterized in comparability to either matrix, or traditional composites, commonly referred to as "particle microcomponents," by enhanced thermal, mechanical, barrier, fire-retardant, and optical features because their extraordinary layering or exfoliation stage morphology optimizes the interface between organic and inorganic phases. CNTs have been widely studied in the fields of chemology, physics, materials science, and electrical engineering, since the discovery of carbon nanotubes (CNTs) in Iijima in 1991 [2]. Carbon nanotubes, with a low mass density and a large aspect ratio (usually around 300–1,000), are very versatile. They have an outstanding combination of mechanical, electrical, and thermal properties, which enable them to replace or supplement traditional nanofilters with excellent candidates for producing multifunctional polymer nanocomposites. Several studies have also shown the importance of the analysis of thermal properties, particularly TGA, of CNTS-containing nanocomposites, as the heat stability of the nanotube-filled polymer matrix has been significantly increased in comparison with the unfilled nanotubes [4]. With regard to carbon nanotubes, PNs generally use X-ray diffraction (XRD), electronic transmission microscopy (TEM), NMR, and real-space observation as an effective measurement for the intercalation polymers in lamellar galleries. While XRD is an excellent means of determining the interlayer separation of silicate layers from interlaced nanocomposites, nothing can be said about the spatial distribution of silicate layers or structural non-homogeneities of nanocomposites. However, TEM takes a long time and provides qualitative information only on the whole of the sample because of the limited scope of research. For the analysis of many phenomena that occur during the thermal scan of nanofilters and PNs such as melting, crystallization, kinetics, and glass transition, differential calorimeter scanning (DSC) has been widely used. When nanoscale dispersion is achieved, these properties present a strange change. Dynamic mechanical thermal analysis (DMTA) has been used widely as a measure of stiffness, energy loss, and temperature-dependent measurement in nanocomposites [3–6]. The degree and the scale of the dispersion of nanofilter strongly influence DMTA results. The thermal stability of the polymer has been studied using thermogravimetric analysis (TGA) for the effect of the incorporation of nanofilters into the polymer matrix. Vyazovkin, in particular, wrote detailed thermal analysis literature papers. This chapter and many of his own researches provide an important insight into the use of TA methods to study the properties, including limitations, of polymer nanocomposites. This chapter is aimed

Characterization Tools and Techniques

at demonstrating flexible applications in nanomaterial science in the emerging field. In fact, even on nanoscales, thermal analysis can detail the average nanocomposite structure by analyzing large volume samples of many milligram sizes.

5.2 CHARACTERIZATION TECHNIQUES

5.2.1 Diffraction

5.2.1.1 X-Ray Diffraction

The many contributions to this work include nanoclusters, partially crystalline and partially amorphous polymers/fibers, nanomaterial or nanostructures, and protein-coated nanoparticles. Atoms in both of these substance types are organized into periodic arrays (nanocracy) or may be found in random (amorphous) clusters. X-rays can be studied on a range of dimensions, from the atomic nucleus to the nanoscale to the mesoscale (hundreds of nanometers), as can be seen in recent nanocharacterization reviews/books and X-ray diffraction nanostructures. A nanocrystal is a crystal bounded by a periodic atomic order to a nanoscale area of three, two, or one dimension [7,8]. This region is a nano-domain that may or may not fit the nanocrystal dimension. Set in hierarchical and complex patterns, nanomaterials can have atomic order (depending on the location of the atoms in the nanocrystals) or nano–mesoscale (depending on the place of the nanocrystals in the auto-assembly). Usually, a self-assembled nanostructure has two essential dimensions: the height of the nanocrystal and the length of the nano-assembly. Scattering of X-rays and diffraction of Bragg will obtain morphological and structural information of the nanomaterials examined. Figure 5.1 shows a schematic of X-ray diffraction by Bragg's law [9].

Structure of atomic crystal: atomic location/symmetry within the unit cell, unit cell length, and nanocrystalline domain shape/size;
Crystalline mix: crystalline phase description and quantitative weight fraction determination;
Nanoparticle assembly: nanoparticle/nanocrystal position/symmetry in assembly and montage extension.

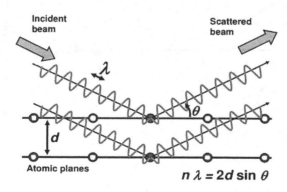

FIGURE 5.1 Schematic of X-ray diffraction by Bragg's law.

The crystal state and crystal lattice were the main parameters until the X-rays and diffraction had been thoroughly understood. A periodic sequence of atoms is made up of the crystalline matrix. There may be different kinds of X-ray interactions when the crystalline material comes in contact. Electrons that transmit nucleus scattering are practically insignificant in elastic or inelastic dispersion. If the energy of a photon that enters and exits is the same, it is called elastic dispersion or Thompson dispersion. The radiation released is dealt with positively or negatively. It checks for positive interference and analyzes diffraction peaks. The Bragg rule obediently intervenes, but it does not do so destructively. The mechanism is shown schematically in Figure 5.1. At an angle θ, the beam is exposed to an X-ray wavelength ray with its tangential base. This term can be explained in the form of mathematics in the Bragg law:

$$n\lambda = 2d\sin\theta \qquad (5.1)$$

where λ is the wavelength, d is the path difference, θ is the incident angle, and n is an integer.

Two XRD methods are mostly used: One is the Bragg law and the other one is the Laue method. Figure 5.2 shows a schematic of X-ray diffraction by the Laue method.

In both cases, diffracted strength of the X-ray beam is measured against the angle 2θ of diffraction that gives the pattern of the material. The XRD model shows sharp maxima (peaks) for the crystalline materials, and while those peaks are not used for the amorphous solids, a large maximum (hump) is shown. Miller (hkl) indices define the various crystal planes in a crystal. These are the three integral numbers associated with the mutual intersection values of a given plane with the cell-unit axes. The diffraction pattern appears in a variety of spots as the material is irradiated into a single crystal. These diffraction spots can become circles when the substance is in a polycrystalline area. Since polymers are not completely amorphous, they display a degree of

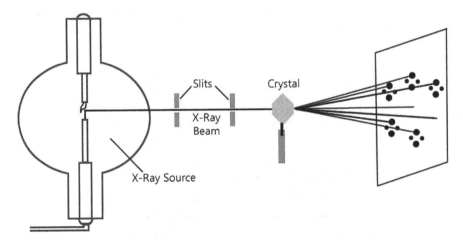

FIGURE 5.2 Schematic of X-ray diffraction by the Laue method.

Characterization Tools and Techniques

crystallinity that tests how crystalline the material is. They convey the crystallinity. The intensity of the diffracted beam in the crystalline component (IC) and the intensity in the amorphous portion of the diffracted beam are divided into two parts.

To predict properties and to evaluate the potential for application in different fields, knowledge and understanding of the degree of crystallinity in polymers is essential.

5.2.1.1.1 X-Ray Diffraction Applications

This approach is commonly used to determine quantitative analysis, phase recognition, and structure of imperfection. Through applying this technique correctly, detailed descriptions of the structural materials can be obtained.

- Determination of the number of unit cells
- Crystalline structures
- Crystalline sample analysis
- Description of minerals
- Quantitative mineral analysis
- Thin-film sample characterization
- Texture characteristics
- Identification of missing lattices
- Evaluating the density of dislocation
- Mathematical analysis of the relative orientations of the crystals Facilities
- Facility of instrument availability
- Simple analysis of data
- Flexible method for uncertain mineral detection
- Unambiguous mineral determination in many instances
- Limited preparation of samples
- Nondestructive
- Evaluating mixed phases
- Fast and quick preparation of samples
- Simple data analysis as these units are widely available Boundaries
- Amorphous substance classification by WXRD being not feasible.
- Depth-profile information missing.
- Difficulty of indexing patterns for non-isometric crystal systems.
- Detection limit for single-phase content being 5% by weight.
- Possibility of peak overlay.

5.2.1.2 Neutron Diffraction

The neutron diffractometer's angular resolution is typically a little lower than the X-ray diffractometer. Neutron sources have been converted into different perspective devices during the last two decades of spallation with a significantly improved angular solution [8]. A fully transformed lath martensite specimen was in situ deformed by tension in a high-resolution and high-intensity neutron diffractometer. The peaks of diffraction are perfectly symmetrical. In fact, during tensile deformation the peaks of diffraction become asymmetrical. The peaks with diffraction vectors are parallel to and perpendicular to the axis of the traction with a longer tail in the direction of small and wide d^*, where d^* is the reciprocal space coordinate. Usually, asymmetrical peak enlargement indicates the existence of long-range internal stresses between coherent grain areas [9,10]. These bundles, either parallel to or oblique to the larger Schmid component, are lath bundles connected with lath planes. The dislocations of lath packets easily glide on longer medium-free paths parallel to the biggest Schmid element. Dislocations in lath-oriented packets oblique to the largest factor Schmid at the lath borders are blocked, which may be accompanied by short mid-free routes. Figure 5.3 shows a schematic of neutron diffraction. Due to the fundamental properties of neutrons as free-state particles, i.e., beyond the atom nucleus, experimental techniques focused on neutron dispersal are of particular interest in research on new materials. It is reasonably stable and has an average lifetime of about 14 min; it does not have a net electric charge as described in its own name – i.e., it is neutral; and it does not have an electric dipole time, but has a magnetic dipole moment linked to a magnetic cornal momentum or a spin. Usually, they are poorly absorbed as opposed to other charged particles, or even X and gamma images, by direct contact with matter. The main interaction is with the nuclei [11].

It is not clear how neutrons and nuclei interact, but Fermi's pseudo-potential theoretically is not understood. This interaction occurs in an elastic system in which the neutron's kinetic energy is held in contact with the nucleus. Typically, we call neutron diffraction the elastic mechanism of dispersion, and wave bases including notions such as constructive and destructive interference can show the mutual neutron-dispersing effect of a group of atoms. Neutron diffraction experiments are used extensively to build and refine structures in crystalline form. They are applied by experimental material methods and analytical techniques, which are very close to the techniques of the X-ray diffraction in the form of monocrystals or polycrystals (powder). With the Rietveld method, the neutron diffraction and X-ray diffraction data can be processed on the same network, typically on the same algorithm, for the structural refining programs of polycrystalline materials such as FullProf and GSAS. Shortly after Chadwick had discovered the neutron in 1932, the first experiments were performed in the same way as an electron or X-ray wave on a probability of neutron beam crystal-diffraction. The comparison is clear and logical. Neutrons are very large hydrogen mass particles that have an excellent potential to penetrate solid matter, since they do not have an electrical net charge [12–15]. According to quantum mechanics, each particle of momentum $p = mv$ (cm – mass; v – rapidity) is a µl wave of the wavelength of the relationship between de Broglie, $\mu = hp$, where h is the Planck constant. The average wavelength of the Kα doublet of a copper anode X-ray tube is closely approximated by that value, commonly used for the purposes of structural analysis of different materials in polycrystal diffractometers. The first signs

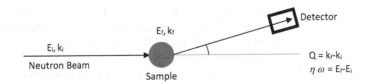

FIGURE 5.3 Schematic of neutron diffraction.

of neutron diffraction [3–5] appeared in the 1930s from low-intensity experimental assemblies. The use of neutron diffraction as an accessory technique in crystallography in the structural characterization of materials during the postwar years had its main impetus in the advent of nuclear reactors and particle accelerators. The latest sources will produce beams with abundant thermal neutrons. The neutron beams are made up of a spectrum of cinematic energy, a "hot" radiation in wavy terms. We can produce a monochromatic beam from a white neutron beam by using monocrystals, much as we can with X-rays. The microscopic magnetic structure of a substance is thus exposed by neutron diffraction. Magnetic dispersion requires an atomic shape factor since the cloud of electrons that is much larger around the small nucleus triggers it. Therefore, the magnetic contribution amplitude to the peaks of diffraction would decline to higher angles. Using neutrons as tools for the study gives scientists a unique insight into the structure and properties of materials that are significant in biology, physics, and engineering. The dispersion of neutrons is a simple example of where and what the atoms do. It enables researchers to see, in real time, how the structure of the material varies with temperature and pressure and in the magnetic or electronic field. It also measures the atomic movements of the electron, which offers important energy-saving information for materials such as magnetism or the ability to control electrical energy. Neutrons have wavelengths of 0.1–1,000 Å and are allowed to detect other radiation types [16–19]. These unique characteristics allow them to supply us with information that is often impossible to accomplish using other methods because of the neutrons to behave as particles or as magnetic microscopic dipoles. For example, the neutrons that disperse gases, liquids, and solids give us details about the structure and magnetism of these materials. Neutrons are nondestructive radiations that can penetrate into matter profoundly. They are ideal for use under extreme strain, temperature, and magnetic field conditions in biological materials and samples [20–22]. Moreover, atoms of hydrogen are neutron-sensitive. It is an effective instrument for the study of hydrogen, organic molecular content, biomolecular sample, or polymer storage materials.

Certain neutron applications in various fields of research are as follows:
Physics of condensed matter, chemistry, and study of materials:

- Clarity in processes such as electric battery charging also of unexplained phenomena.
- Metal storage of hydrogen is a significant feature of the production of renewable power.
- Analysis of basic parameters (e.g., elasticity) in polymers (e.g., plastics).
- Colloid analysis provides new insights on different topics, such as oil extraction, cosmetics, medications, foodstuffs, and chemicals.

Biological studies:

- Naturally rich biological materials with hydrogen and other light elements are ideal for studying neutrons.
- Protein cells and membrane

- Nucleic and fundamental particle physics
- Research on physical properties of neutrons and neutrinos
- Incredibly slow output of neutrons down to 5 m/s (neutrons are normally around 2,200 m/s) from the reactor.
- Atomic and nucleus structure fission tests.

Science integrated:

- As nondestructive neutron diffraction is ideal for studying different phenomena of scientific materials.
- Visualization of residual tension in textiles.
- Stiffness and lateral corrosion signs.
- Inhomogeneity of substances and trace elements.

5.2.2 Microscopy

5.2.2.1 Electron Microscopy

5.2.2.1.1 Scanning Electron Microscopy

SEM for material characterization and surface characterization is among the most widely used microscopies. These microscopes are 1–5 nm size. They also have a wide field depth besides high resolution, and thus, the images seem to be three-dimensional. The SEM theory is that electron weapons create and speed up electrons by means of lenses that focus the beam at a very short spot. These electrons interact with the specimen to about 1 µ deep and produce signals for the image. Backscattered electrons, secondary electrons, and X-rays are three most common signs. The distributed electrons are elastically dispersed electrons and contrast the composition of the specimen according to its atomic number. These electrons are highly driven and come from the depth (1 µm or more) of the specimen. Secondary electrons from the top surface of the specimen are energy-effective electrons (a couple of nm) widely used to visualize samples' topography. Cellulose nanocomposites or bionanocomposites produce polymers that are polymeric in both matrix and reinforcement, i.e., non-conductive, and consist of low–atomic number elements that are primarily used in the typographic imaging process of these materials [23–25].

5.2.2.1.2 Transmission Electron Microscopy

The TEM working theory is that an ultra-thin component of the specimen transmits electrons that are also of high energy. When the beam enters the specimen, the image is created with electron dispersion. Filament produces electrons. Underneath the electron fuselage, there are two or three condenser lens that demagnify the fire arm's beam and control its diameter when they are mounted in the target lens just under the condenser lens. There are two lenses left after the objective lens: the intermediate lens and the lens of the projector. Every image generates a real, enlarged picture that creates a fluorescent picture on the film or screen. The difference is due to electron dispersion in the TEM picture. Bright field (BF) is the mode of imaging in which an objective opening is positioned in order to create the image with directly dispersed electrons [26–28]. In the specimen, regions that are more thick or densely expanded

Characterization Tools and Techniques

appear darker in the image, because the objective aperture prevents the electrons from being very dispersed. In BF, white is a picture area without a reference. There are three basic mechanisms of contrast in TEM that can result in image creation:

- Diffraction against comparison,
- Mass thickness comparison, and
- Contracting.

Cellulose nanocomposites consist of elements with small atomic numbers and thus weakly disperse electrons and provide a lack of TEM contrast. For any material, the mechanism of mass contrast can be manipulated by deliberately staining the thin specimen with heavy metal, which emphasizes special characteristics. Uranium acetate is, for example, an important staining agent for contrast enhancement of CNs. The characteristics of particularly enhanced nanocomposites are affected directly by the distribution within the nanoparticle process matrix. The distribution of the nanoparticles along a surface or if they are embedded in a polymer matrix must be described as this relates to mechanisms for product enhancement for the resulting composite properties. Transmission electron microscopy is used to examine a nanocomposite material's think thickness and thus, due to CN distribution and dispersion, is the most effective microscopy technique [27–29].

5.2.2.1.3 Atomic Force Microscopy

Atomic force microscopy is a microscopy scanning method for scanning the surface properties of nanometers and is used to classify nanocomposites. The topographical imaging is typically completed by scanning the composite surface sample (AFM, typical 10 nm curvature radius) by controlling the reaction of the sample to the interaction of the sample surface. Two operating modes (contact and intermittent contact) can be used, and under various conditions (vacuum, steam, fluid), experiments can be carried out. The touch mode scans an AFM sample tip over the surface to keep the input between the tip and the sample secure [30–32]. The AFM probe is vibrated at its frequency of resonance in the intermittent touch mode (or "tapping" mode), and feedback is used to retain the force in certain aspects of the vibration of a probe (such as amplithium). Intermittent contact mode has lower lateral forces than contact mode and offers additional contrast channels, for example, a phase picture. AFM topography imagery of nanocomposites has been used to depict surface roughness. Figure 5.4 shows a schematic of atomic force microscopy. The quantitative resolution of the sub-nanometer and the qualitative resolution of the nanometer laterally permitted comparative studies of the treatment of the robustness and ending of the nanocomposite cellulose surface.

The CNC alignment within nanocomposites has been identified by AFM imagery (topography, phase imaging, etc.) where the CNC alignment degree can be quality-defined using various image analysis methods. The raw image in the AFM is usually post-processed for highlighting specific particles, with some functions for defining the CNC's long axis. It determines the angle of the CNC's long axis to a given axis (e.g., sample geometry, or alignment direction of the process). A histogram that sums up the percentage of CNCs from 0° to 90° is produced for a certain angle. Notice that this represents a two-dimensional CNC alignment because all CNCs are parallel to an AFM image

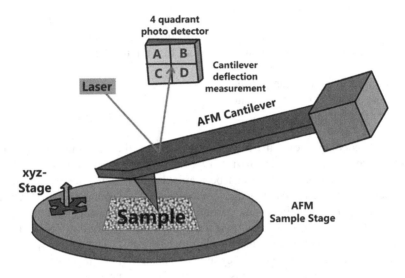

FIGURE 5.4 Schematic of atomic force microscopy.

plane. For CNC composites, this additional degree of freedom with CNC orientation may also be used for the design of three-sized arrangements of short fiber composites using the form of the elliptical cross section type on a polished surface [32–35].

The EM's importance in science and technology of nanocomposites derives primarily from two factors. Next, electron microscopic information coming from other sources is even more accurate than that. Using electron microscopy, details about the morphology of the polymer matrix as well as the filler and adhesion between them can be measured with nanometer resolution at the same time. Second, electron microscopy enables the analysis of the reaction of all of the composite's structural information to the applied load (sometimes even in situ), allowing the creation of tailored material. Electron microscopy is the only technique that offers very clear proof of intercalation and exfoliation of the filler in the polymer matrix, allowing the morphological characteristics of the polymer nanocomposites to be quantified straightaway.

5.2.3 Spectroscopy

5.2.3.1 FTIR Spectroscopy

Fourier transform infrared spectroscopy (FTIR) is a measuring technique used to classify molecular compounds by infrared radiation. When the sample is irradiated by infrared, it absorbs radiation which causes vibration of the chemical bonds in the sample. Created infrared spectrum requires the existence of chemical bond. Regardless of the molecule's structure, functional groups absorb infrared radiation at a particular wave number. A look at the particular absorption peak in the infrared spectrum will recognize the presence of various functional groups. Interferometer consists of one beam splitter that is capable of separating the incoming infrared radiation into two rays. The beam splitter transmits part of the beam, and the other part is mirrored. The two split beams hit two mirrors, of which one is fixed and the other is mobile, and then recombine at the beam splitter after a difference in direction has been made [36]. This beam then traverses the sample before reaching the detector. The role of moving mirror is to generate interference by changing the distance traveled by one beam between two reflected beams. The strength of each beam traveling to the detector and returning to the source depends on the difference in the direction of the beams in the interferometer's two arms, due to the interference. The variance of the intensity of the beams that travel to the detector and return to the source as a function of the difference in the optical path gives a spectrum called interferogram. An interferogram is the sum of sinusoidal waves with a range of wavelengths and is converted to a single infrared spectrum using Fourier transformation. Fourier transformation is a mathematical method that transfers information between a function in the time (t) domain and its corresponding frequency domain and is given as an interferogram.

$$F(\omega) = \left(\frac{1}{\sqrt{2\pi}} \int f(t) \exp(-i\omega t) dt\right) \quad (5.2)$$

This spectrometer is able to collect spectra in the mid-IR, far-IR, and near-IR spectral ranges. In this frequency range, absorption peaks due to organic compounds are readily observable.

5.2.3.2 UV–Visible Absorption Spectroscopy

Ultraviolet–visible spectroscopy (UV–Vis) is an optical characterization method used to determine the absorption of photons in a sample in the spectral range of the ultraviolet. This technique tests electronic transitions from the ground to the excited state and is complementary to the spectroscopy of fluorescence. SHIMADZU UV-2450 spectrophotometer is being used mainly to measure the nanoparticles' absorption and transmittance, scintillating dye, and nanocomposite scintillators. In general, there are a UV–Vis spectrophotometer, an electromagnetic source of radiation, a diffraction grating, a sample cell, and a detector [37]. The entire UV–Vis spectral range is filled by a combination of a deuterium lamp for the UV region of the spectrum and tungsten or halogen lamp for the visible area. Light from the source of UV–V is light is separated by a diffraction grating and split into its component wavelengths. Figure 5.5 shows a schematic of UV–visible absorption spectroscopy.

The radiation of only one selected wavelength exits the monochromator through the exit slit, by shifting the dispersing element or the exit slit. Then, a beam splitter splits the selected wavelength light into two separate directions, where one of the light beams passes through the sample and the other passes through the reference sample in the sample cell. The reference sample only contains the solvent that is used to disperse the sample. The two beams of light are then gathered and measured using a detector. Usually, the detector is a photomultiplier tube, a photodiode, a photodiode array, or a charging-coupled device (CCD). The intensity of the sample cell (I), the reference cell intensity (I_0), and the absorption (A) for the given wavelength are correlated with the Beer–Lambert law:

$$A = -\log\left(\frac{I}{I_0}\right) \quad (5.3)$$

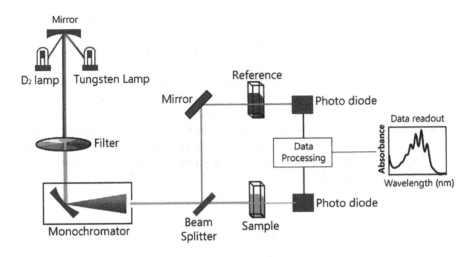

FIGURE 5.5 Schematic of UV–visible absorption spectroscopy [37].

The absorbance measured corresponds to a given wavelength, and the UV–Vis spectrophotometer repeats this process for the whole range of selected wavelengths to provide a spectrum of absorbance vs. wavelength. For accurate absorption calculation, it is important that the sample concentration is low to prevent any inconsistencies due to light dispersion.

5.2.3.3 Raman Spectroscopy

Raman spectroscopy is a method used most generally for the determination of molecular modes of vibration, though it is possible to detect rotational and other low-frequency structural modes, named after the Indian physicist C. V. Raman. Raman spectroscopy in chemistry is widely used for structural molecular recognition fingerprinting.

Raman is primarily dependent on inelastic dispersion of photons, known as Raman dispersal. A monochromatic light source, typically within the visible, near-infrared, or near-ultraviolet laser spectrum, may be used, though X-rays can be used too. Laser light interacts with molecular vibrations, phonons, or other device excitations to allow laser photons to transfer their power up or down. The energy change provides information on the device's vibrational modes. Infrared spectroscopy generally gives closer, additional data. Normally, a laser beam illuminates a sample. An electromagnetic radiation received by a lens is transmitted by a monochromator from the illuminated stage. A filter or band-pass filter is filtered at an overhead rate, while the remaining filter is dispersed over the laser line-scattering detector.

Spontaneous Raman dispersal is generally very weak, and so for many years the main challenge to obtain spectrum from Raman was to isolate the light of the powerful laser light (called "laser refusal"), which Rayleigh dispersed. Spectrometers of Raman also use holographic gratings and different stages of dispersion to attain a high degree of laser repression. In the past, photomultipliers were the chosen detectors for dispersive Raman systems, leading to long buying times. Nevertheless, modern instrumentation makes the use of the notch or edge filters almost universal.

Characterization Tools and Techniques

In combination with CCD detectors (AT or Czerny–Turner (CT) monochromators, they are most widely used as single-stage dispersive spectrometers and, in NIR lasers, as the Fourier (FT) spectrometers [37].

The term "Raman spectroscopy" usually refers to Raman that is not absorbed into the sample by the laser wavelength. Raman has a range of other versions: Raman enhanced base, Raman spatially offset, Raman enhanced edge, Raman polarized, Raman stimulated and super-Raman spatially offset.

The Raman effect's frequency in a molecule corresponds to the electron's polarization. It is a kind of inelastic light dispersion that activates the sample with a photon. For a short time, before releasing the photon, this excitement keeps the molecule in simulated energy environments. Inelastic dispersion results in less or more energy from the emitted photon than the intensity of the photon incident. After the dispersion phase, the sample is in a different rotating or vibrating state.

After moving the molecule to a new rovibronic (rotary vibrational electronic) state, the broken photons move in a different direction and thus at a different frequency to keep the total system energy constant. The energy differential is proportional to the difference between the original and final states of the molecule. If the final state is higher in energy than the initial one, the transmitted photon is transferred to a lower frequency (low energy) to preserve overall energy [38]. Figure 5.6 shows the absorption and scattering in Raman spectra. Stokes shift or downshift is referred to as frequency transition. If the final condition was less energy, the transmitted photon would be moved to a higher frequency called an anti-shock transition or upshift.

The polarization of its electric dipole–electric dipole must adjust when it comes to vibrational coordinates corresponding to the rovibronic state in order for a Raman effect on a molecule. Raman's amplitude is commensurate with the difference in polarity. Thus, the Raman spectra depend on the rovibronic state of the molecule (dispersive force according to the frequency shifts).

The Raman effect is based on the interaction between the electron cloud of the sample and the monochromatic external electrical field, which may cause dipole moment inductions depending on its polarization. Although the laser light does not reach the atom, there can be no obvious difference between the energy ranges. The Raman effect

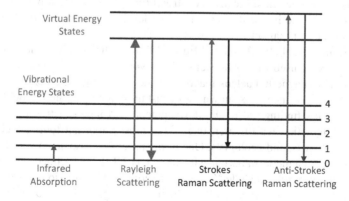

FIGURE 5.6 Raman spectra.

could not be correlated with emissions (fluorescence or phosphorescence) when a molecule in an excited electronic state emits a photon and, in many situations, returns in an excited electronic state. Raman dispersion is similar to the infrared (IR) absorption in which the frequency of the absorbed photon corresponds to the frequency difference between the original and final rovibronic state. Raman's reliance on electric polarization by dipole electrical derivative is frequently dependent on the electric dipole moment derivative suitable for IR spectroscopy. This contrasting function permits the examination of rovibronic transformations not subject to IR spectroscopy, as seen in a centrosymmetric molecular law of mutual exclusion. Strong Raman transitions of power have weak IR as well, and vice versa [39]. A small length shift such as this, when a bond is extremely polarized, has no effect on polarization. Therefore, Raman's relatively weak dispersers are vibrations involving polar bindings. However, such polarized bonds hold their electrocharges (unless neutralized by symmetry factors), which lead to a greater change during vibration of the net dipole moment and to a strong IR unit. Furthermore, relative neutral relations undergo significant polarization variations in a vibration. The moment of the dipole, however, is not affected similarly, so although Raman scatterers are often strong vibrations involving this form of relation, they are faint in IR. A third technique of vibrational spectroscopy, inelastic incoherent neutron spreading (IINS), may be used to determine the vibration frequency of highly symmetric molecules in IR and Raman, both of which may be inactive. The rules for IINS selecting or permissible transitions differ from those for IR and Raman, as the three strategies are mutually complementary. The relative intensities have different details for photons for IR and Raman and the neutrons for IIS, however, given the various ways of interaction between the molecule and the incoming particles.

5.2.3.4 Nuclear Magnetic Resonance

There is a large market for nanomaterials to develop characterization techniques. Several biophysical techniques have contributed significantly to the resolution of the problems associated with their characterization; however, they lack the resolution of the atomic stage. Due to the presence of nanomaterials in different permutations and combinations, containing different combinations of nanomaterials and ligands for different applications, nuclear magnetic resonance spectroscopy (NMR) can irrevocably analyze the diversified spectrum of structural and chemical properties of nanomaterials in solid and liquid states with atomic-level resolution. Figure 5.7 shows a schematic of nuclear magnetic resonance.

At the same time, Felix Bloch and Edward Purcell discovered nuclear magnetic resonance in condensed matter at Stanford in 1946 using various instruments and techniques. The magnetic nucleus reaction in the same magnetic field, as the field is tuned into a resonance, was shown in a constant magnetic radiation field in both groups [40]. This finding led to the development of a new spectroscopy method, radioactive interaction with matter, one of the most important tools for chemists, physicists, biologists, and geologists. One of the processes associated with the interaction of electromagnetic radiation with matter is NMR, as defined by scientists. Other common examples include X-ray attenuation, visible emission and absorption rates, food microwave heating, and RF induction heating. RF is also a typical example of this type of interaction. [12].

Characterization Tools and Techniques

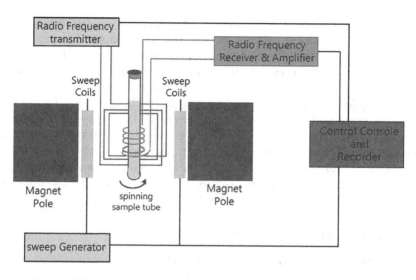

FIGURE 5.7 Schematic of nuclear magnetic resonance.

The radiographic X-rays have a standard frequency of approximately 1,018 Hz and an orange light of approximately 5×10^{14} Hz has a sodium lamp. Microwave cookers operate at a frequency of 10^9 Hz, while RF heaters use a frequency of 10^6 Hz. The experimental methods used to process and track radiation are substantially different in each of these cases. But the theoretical understanding of the processes involved also includes similarities. Quantum theory demonstrates these effects as far as transitions between various energy states are concerned. The frequency of radiation is connected by Einstein's relation to the energy differential M (= E2 – EI): M = hV, where h is the Planck constant. NMR is a phenomenon that happens when a second magnetic oscillating field is immersed in atoms' nuclei. Some nuclei have this effect, while others do not, depending on their own spin property. Table 5.1 provides a number of nuclei with spin. It is necessary to note that we perform atomic nucleus tests with NMR, not electrons. NMR spectroscopy net spin linked to protons and neutrons (both of which are 1/2 spin quantities) and positive charge distribution are two properties of a nuclear particle essential to NMR spectroscopy understanding.

TABLE 5.1
Number of Nuclei with Spin

Nuclei	Unpaired Protons	Unpaired Neutrons	Spin
^1H	1	0	½
^2H	1	1	1
^{31}P	0	1	½
^{23}Na	2	1	3/2
^{14}N	1	1	1
^{13}C	0	1	½
^{19}F	0	1	½

Spin I = 1/2 nuclei are most widely used in NMR spectroscopy. No spectra can be obtained from I = 0 nuclei, and only spectra from I21 nuclei can occur in specific cases. In the case that I = 1/2 nuclei, two values show that the magnetic momentum vector is optimal for the corner of the nucleus, mI = +1/2 or −1/2, in an external magnetic field. The value +1/2 refers to the relation between the vector and the magnetic field applied, −1/2 against it. The values for Mp are I, (II) [14–19]. The (−I + 1), I. For I = 1, m is equally equivalent to the alignments of +1, 0, and −1, perpendicular to and opposite the ground. In the absence of a magnetic field, the two atomic orientations deteriorate. However, this degeneration could impede the life of an external force. In the case of a nucleus with I = 1/2, the mI = +1/2 is less force and greater than −1/2. This removes the excess nucleus at the lower energy level, and the two states are nearly filled at average temperatures.

5.2.4 THERMAL MECHANICAL ANALYSIS (TMA)

5.2.4.1 Differential Scanning Calorimetry (DSC)

Differential scanning calorimetry of the heat of organoclays, polymer/stick nanocomposites, or nanotubes, including the transformational glass (Tg), melting, crystallization, and treatment, was commonly used in investigating various phenomena occurring during thermal heating. DSC is one of the most common methods used to study the transition from the Brownian movement of main polymers and their composites. The chains are related to the transition from glass to rubbery and the relaxation of dipole. Figure 5.8 shows a schematic of differential scanning calorimetry (DSC).The DSC technique, with reference to clay nanocomposites, shows substantial changes in Tg resulting from the incorporation into multipolymers of nano-sized montmorillonite [41–42]. This impact was due generally to the containment in the silicate gallery of intercalated polymers that prevents segmentary movements of the polymer chains. Changes in the glass transition temperature were also interpreted for the polyurethane (PU) urea nanocomposites due to the active bonds between the polymer chain and the silicate

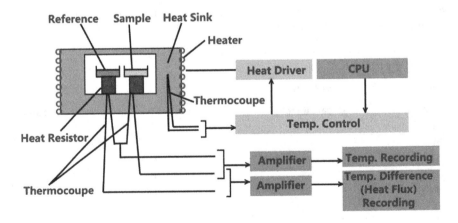

FIGURE 5.8 Schematic of differential scanning calorimetry (DSC).

surface. The attached polymer chains can become a region in which the rest component is less than in bulk. The nanocomposite polymer with intercalated and exfoliated silicone has a restricted relaxing behavior that depends principally on exfoliation and interacting strength of silicate surfaces and macromolecules. Thermogravimetric research on different PNs demonstrated an enhanced thermal stability (i.e., higher thermal degradation temperatures), including poly(methyl methacrylate) (PMMA), poly(dimethyl siloxane) (PDMS), polyamide (PA), and polypropylene (PPP), in certain polymers with the aid of montmorillonite and carbon nanotubes. In polymer clay nanocomposites, it is widely agreed that the increased thermal stability of the PN is primarily due to the development of a char that prevents the spread of volatile decomposition products as a consequence of the decline in permeability normally seen in nanocomposites that are exfoliated. However, there is no existing established mechanism for the exact deterioration of clay, which may be associated with morphological changes in the loading of exfoliated or intercalated animals on the clay. Exfoliation prevails at low clay loadings (approx. 1 wt.%), but nanoclaying is not sufficient to improve thermal stability in the formulation of residue. In addition, clay can slow oxygen diffusion in the air and cause thermooxidative reactions in the atmosphere [43].

The effect of clay on the thermal stability of nitrogen depends on the method, on the other hand; therefore, there is no empirical evidence that a reduction in permeability improves the thermal stability. With rising levels of clay (2–4 wt.%), much more exfoliated clay is more efficiently produced, thus promoting the thermal stability of the nanocomposite. The intercalated structure is the dominant population at an even higher level of clay charging (up to 10 wt%). While char is produced in large amounts, the different morphology of nanocomposites is probably not sufficient for maintaining high thermal stability. However, in the chemical structure of the polymers, the form, and the direction of modification of clays, PN's degradation behavior is known to play an important part. However, the formation of carcass may not have affected thermal stability, since this is done at the very end of decay. The heat and degradation processes of nanocomposites are studied by two important studies using different polymer matrices. The topic addresses fundamental modifications in thermal actions of the various polymer matrices: polyolefins, polyamides (PAs), polymer container styrene, poly(vinyl chloride) (PVC), polymethyl methacrylate (PMMA), polyimides (PIs), epoxy resins, polyesters, polyurethane (PU), and ethylene-propylene-dioterpolymer (EPDM) [20–24]. The "labyrinth" effect limits the distribution of oxygen in the nanocomposite during thermal degradation. Similarly, MMT layers inhibit the distribution of gases released during degradation in samples exposed to high temperatures, thus helping to keep the polymer in contact with the non-oxidizing environment. In addition, the thermal conduction of MMT layers is expected to minimize. The movements of polymer chains are constrained in the presence of MMT layers that are highly interactive with the polymer matrix as seen in the previous section on OMMT-induced Tg changes. This influence provides enhanced stability for polymer/MMT nanocomposites. It is also proposed that the chemical interaction between the polymer matrix and the surface of the tile sheet during thermal degradation could help improve carbon output in the thermal degradation process. Some scientists have demonstrated that the catalytic activity of nanodispersed clay induces a CarC reaction. Nanodissipated MMT layers often interact in a manner that strengthens macrochain frames and restricts thermal movements in polymer domains with polymers. In general, thermal stabilization and dispersion of organoclays in polymer nanocomposites containing MMT are involved. The synthesis methods affect the thermal stability of nanocomposite polymer/MMT, provided they control the degree of dispersion of the clay layer. Extensive research is currently being conducted on the synthesis of new thermally stable (including oligomeric) modifiers capable of ensuring good compatibility and enhancing thermal stability of nanocomposite due to low migration properties. In contrast to thermogravimetric research, several groups reported improved thermal stability in the composites of nanotubes/ polymers. Polymers similar to nanotubes can be slowly degraded to permit Tpeak to pass to higher temperatures [31,32]. The improved thermal stability in nanotube/ polymer composites, which makes heating dissipation inside the composite possible, can also be due to another possible mechanism. The increased thermal stability observed indicates that nanotubes can be helpful in polymer matrices as fire retardants.

5.2.4.2 Thermogravimetric Analysis (TGA)

The technique of thermogravimetry (TGA) tests the weight change in a sample whether it is heated, refreshed, or kept at constant temperature. The main aim is to distinguish the materials by their composition. Plastics, elastomers, and thermosets, mineral compounds and ceramics, and a wide variety of studies are the application fields in the chemical and pharmaceutical industries [9,10]. The sensor consists of six thermocouples, in addition to weight adjustment, if you want to calculate simultaneously the heat flow (DSC), and the heat flow decides the estimated or measured temperature difference TGA/DSC [47].

The characteristics and capabilities are given as follows:

- Resolution of ultramicrograms over the whole measurement scale
- Testing sample masses and quantities, small and large
- Evaluating the atmospheric samples up to 1,100°C
- Depending on a leader in integrating technology
- Detecting the thermal events simultaneously
- Making sure the measuring environment is well established COMMUNITIES
- Adsorbing and desorbing the gases
- Study of the quantitative material (humidity, fillers, polymer materials, etc.)
- Description of the products for decomposition
- Stable thermal conditions
- Crystallization, melting behavior, and glass transitions
- Heat power

5.2.4.3 Dynamic Mechanical Thermal Analysis (DMTA)

The DMTA is commonly used in the analysis of nanocomposites because it tests the strength of the material for the recovery or maintenance of the mechanical force and the failure module (E') of two separate nanocomposite components. In general, DMTA data showed significant stock module improvements over a broad range of temperatures, including PVDF, PP, and PMMA, for a wide range of nanocomposites with MMT [44–45].

In determining the thermal expansion (TEC) coefficient of PA-6-based nanocomposite materials PP, PA, and PS, TMA is highly sensitive. In general, CTE has been found to be lower in nanomaterials in relation to unmodified polymers, particularly for low OMMT levels. In addition, it was used to evaluate the expansion and contraction of cross-linked or filled materials such as nanocomposites. The latter result suggested that exfoliated platelets should not be uniformly placed around FD because the ideal configuration of the disk platelets in an isotropic matrix would generate the same expansion coefficients for both FD and TD. This can contribute to lower thermal expansion in an FD chain than in a TD chain. Of course, the differences in the orientation of polymer crystallite may also vary in both directions. The anisotropy and the platelet orientation effects may explain this trend. The results of TMA indirectly provide information on the nanocomposite layers of MMT's spatial orientation. The above findings were similarly determined by TMA measurements on multiwalled carbon nanotubes (MWNTs), infused between and through glazed fibers, along the thickness direction. In the development of multiscale epoxy composites reinforced by glass fiber, both pure and functionalized MWNTs were used [46].

Polymer nanocomposites (PNCs) are materials consisting of a polymer matrix with embedded particles 100 nm or smaller in size. Typical nanoparticles are carbon nanotubes, or nanofibers, graphenes, and nanoclays.

Polymer nanocomposites exhibit improved properties than unfilled polymers, making them desirable for a range of technological uses. Particularly, the desired properties are the greater mechanical resistance of polymeric materials and low weight. In addition, the integration of nanocomponents can lead to increased chemical and heat resistance, as well as electrical conductivity. Polymer nanocomposites are widely commonly used in the aviation and automotive industries, as well as in windmill blade construction materials [47].

Mixing the nanoparticles into the molten polymer matrix using extrusion will create polymer nanocomposites. One way to achieve proper mixing during the extrusion process is by using nanoparticles that are predispersed in a carrier liquid and feeding the dispersion into the extruder. The composite material exhibits the desired properties, only when the particles are homogeneously distributed inside the polymer matrix without creating any larger clusters.

The mechanical properties of a polymer nanocomposite can be checked using dynamic mechanical thermal analysis (DMTA). DMTA can be performed with a rotational rheometer in torsion. As the temperature varies continuously, the material is subjected to oscillatory shear. The data obtained are used to determine characteristic phase changes such as melting and crystallization or glass transformation. In addition, DMTA is used to determine the mechanical efficiency of solid materials with significant application-related properties such as fragility, rigidity, impact resistance, or damping. DMTA is used to obtain the rheological parameters such as loss modulus (G''), storage modulus (G'), and the loss or damping factor (tan γ) [23].

The loss modulus describes the viscous properties, and the storage modulus describes a material's elastic properties. The storage module for fluids is smaller than the loss module, and vice versa for solids. The ratio of G'' and G' is the factor of loss and is also a measure of a material's damping properties. Figure 5.1 shows a schematic diagram for DMTA on a semicrystalline polymer. There are various methods to defining the transition from glass. Usually, polymer nanocomposites are in the glassy state at room temperature and exhibit high G' values, which indicates the material's high rigidity. Polymer nanocomposites in the glassy state exhibit higher G' values than unfilled polymers, suggesting their greater mechanical power.

For copolymers and polymers that carry side chains, smaller phase transitions can take place at temperatures far below the main glass transition. The extra peak in the damping factor will improve a polymer's impact resistance. An example of such material will be high-impact polystyrene (HIPS), an engineering plastic with a polystyrene backbone and rubber side chains [48–50].

5.3 CONCLUSION

The unforeseen synergistic properties resulting from both components prompted substantial research attention from polymer nanocomposites (PNs), i.e., polymer composites filled with nanometer inorganic fillers. In PNs, the efficiency of intercalation polymers in lamellars is typically measured by X-ray diffraction (XRD) and/or electronic microscopy (TEM). While large-angle XRD is a convenient method for determining the interlayer spacing of silicate layers in intercalated nanocomposites, little could be said about whether the structural dislocation of silicate or nanocomposite is structurally uniform. TEM, on the other hand, is very time-intensive and provides only qualitative information on the sample as a whole due to a limited field of study. Thermal analysis (TA) is a valuable method for studying a variety of polymer properties and can be applied to PNs, in particular in the case of montmorillonite nanocomposites, in order to gain further insight into their structure. This chapter offers useful examples of application for the characterization of nanocomposite materials for differential scanning calorimetry (DSC), thermogravimetric analysis (TGA), and dynamic mechanical thermal analysis (DMTA). The XRD pattern and SEM revealed that the obtained powders contain a mixture of micro- and nanoparticles. The chemical identity of the compound can verified through FTIR spectrum, through different peaks.

REFERENCES

1. Bhushan, Binay. "Preparation and characterisation of smart polymer-metal nanocomposite: optical and morphological study." *International Journal of Nano and Biomaterials* 7.3(2018): 219–230.
2. McGlashan, Stewart A., and Peter J. Halley. "Preparation and characterisation of biodegradable starch-based nanocomposite materials." *Polymer International* 52.11(2003): 1767–1773.
3. Halim, Khairul Anwar A., Joseph B. Farrell, and James E. Kennedy. "Preparation and characterisation of polyamide 11/montmorillonite (MMT) nanocomposites for use in angioplasty balloon applications." *Materials Chemistry and Physics* 143.1(2013): 336–348.
4. Larkin, Peter. *Infrared and Raman Spectroscopy: Principles and Spectral Interpretation.* Elsevier, 2017.
5. Butler, Holly J., et al. "Using Raman spectroscopy to characterize biological materials." *Nature Protocols* 11.4(2016): 664–687.

6. Bartolomeo, Giovanni Luca, Guillaume Goubert, and Renato Zenobi. "Tip-Enhanced Raman Spectroscopy (TERS) for nanoscale imaging of biological membranes." *Enhanced Spectroscopies and Nanoimaging* 2020(2020): 11468.
7. Hsieh, Wei-Hsien, et al. "Non-isothermal dehydration kinetic study of aspartame hemihydrate using DSC, TGA and DSC-FTIR microspectroscopy." *Asian Journal of Pharmaceutical Sciences* 13.3(2018): 212–219.
8. Rahman, Md Rezaur, Sinin Hamdan, and Josephine Lai Chang Hui. "Differential scanning calorimetry (DSC) and thermogravimetric analysis (TGA) of wood polymer nanocomposites." *MATEC Web of Conferences*. Vol. 87. EDP Sciences, 2017.
9. Trivedi, Mahendra Kumar, et al. "A comprehensive physicochemical, thermal, and spectroscopic characterization of zinc (II) chloride using X-ray diffraction, particle size distribution, differential scanning calorimetry, thermogravimetric analysis/differential thermogravimetric analysis, ultraviolet-visible, and Fourier transform-infrared spectroscopy." *International Journal of Pharmaceutical Investigation* 7.1(2017): 33.
10. Leyva-Porras, César, et al. "Application of differential scanning calorimetry (DSC) and modulated differential scanning calorimetry (MDSC) in food and drug industries." *Polymers* 12.1(2020): 5.
11. Thamer, Ahmed A., Hashim A. Yusr, and Najwa J. Jubier. "TGA, DSC, DTG properties of epoxy polymer nanocomposites by adding hexagonal boron nitride nanoparticles." *Journal of Engineering and Applied Science* 14.2(2019): 567–574.
12. Mukasyan, Alexander S. "DTA/TGA-based methods." *Concise Encyclopedia of Self-Propagating High-Temperature Synthesis*. Elsevier, 2017, 93–95.
13. Mustafa, Wan Azani, et al. "Experimental study of composites material based on thermal analysis." *Journal of Advanced Research in Fluid Mechanics and Thermal Sciences* 43(2018): 37–44.
14. Candan, Zeki, Douglas J. Gardner, and Stephen M. Shaler. "Dynamic mechanical thermal analysis (DMTA) of cellulose nanofibril/nanoclay/pMDI nanocomposites." *Composites Part B: Engineering* 90(2016): 126–132.
15. Costa, C. S. M. F., et al. "Dynamic mechanical thermal analysis of polymer composites reinforced with natural fibers." *Polymer Reviews* 56.2(2016): 362–383.
16. Kumar Sahu, Santosh, et al. "Dynamic mechanical thermal analysis of high density polyethylene reinforced with nanodiamond, carbon nanotube and graphite nanoplatelet." *Materials Science Forum*. Vol. 917. Trans Tech Publications Ltd., 2018.
17. Karingamanna Jayanarayanan, Nanoth Rasana, and Raghvendra Kumar Mishra. "Dynamic mechanical thermal analysis of polymer nanocomposites." *Thermal and Rheological Measurement Techniques for Nanomaterials Characterization* 3(2017): 123.
18. Goldstein, Joseph I., et al. *Scanning Electron Microscopy and X-Ray Microanalysis*. Springer, 2017.
19. Erickson, Harold P. "Structure seen by electron microscopy." *Plasma Fibronectin*. CRC Press, 2020, 31–51.
20. Zhang, Daliang, et al. "Atomic-resolution transmission electron microscopy of electron beam–sensitive crystalline materials." *Science* 359.6376(2018): 675–679.
21. Zuo, Jian Min, and John C.H. Spence. "Advanced transmission electron microscopy." *Advanced Transmission Electron Microscopy*. Springer Science+ Business Media, 2017.
22. Feist, Armin. Next-Generation Ultrafast Transmission Electron Microscopy-Development and Applications. Diss. Niedersächsische Staats-und Universitätsbibliothek Göttingen, 2018.
23. https://www.britannica.com/technology/scanning-electron-microscope.
24. Kim, Byung Hyo, et al. "Liquid-phase transmission electron microscopy for studying colloidal inorganic nanoparticles." *Advanced Materials* 30.4(2018): 1703316.
25. Carter, C. Barry, and David B. Williams, eds. *Transmission Electron Microscopy: Diffraction, Imaging, and Spectrometry*. Springer, 2016.

26. Juffmann, Thomas, et al. "Multi-pass transmission electron microscopy." *Scientific Reports* 7.1(2017): 1–7.
27. https://www.britannica.com/technology/transmission-electron-microscope.
28. Hansen, Thomas Willum, and Jakob Birkedal Wagner. *Controlled Atmosphere Transmission Electron Microscopy.* Springer International Publishing, 2016, 213–235.
29. El Hajraoui, Khalil, et al. "In-situ propagation of Al in germanium nanowires observed by transmission electron microscopy." *European Microscopy Congress 2016: Proceedings.* Weinheim, Germany: Wiley-VCH Verlag GmbH & Co. KGaA, 2016.
30. Liu, Xi, et al. "Environmental transmission electron microscopy (ETEM) studies of single iron nanoparticle carburization in synthesis gas." *ACS Catalysis* 7.7(2017): 4867–4875.
31. Lorenz, Christiane, et al. "Imaging and characterization of engineered nanoparticles in sunscreens by electron microscopy, under wet and dry conditions." *International Journal of Occupational and Environmental Health* 16.4(2010): 406–428.
32. Lorenz, Christiane, et al. "Imaging and characterization of engineered nanoparticles in sunscreens by electron microscopy, under wet and dry conditions." *International Journal of Occupational and Environmental Health* 16.4(2010): 406–428.
33. Shang, Li, Karin Nienhaus, and Gerd Ulrich Nienhaus. "Engineered nanoparticles interacting with cells: size matters." *Journal of Nanobiotechnology* 12.1(2014): 5.
34. Hassellöv, Martin, et al. "Nanoparticle analysis and characterization methodologies in environmental risk assessment of engineered nanoparticles." *Ecotoxicology* 17.5(2008): 344–361.
35. Weinberg, Howard, Anne Galyean, and Michael Leopold. "Evaluating engineered nanoparticles in natural waters." *TrAC Trends in Analytical Chemistry* 30.1(2011): 72–83.
36. Mishra, Deepti, et al. "Synthesis and characterization of iron oxide nanoparticles by solvothermal method." *Protection of Metals and Physical Chemistry of Surfaces* 50.5(2014): 628–631.
37. Mishra, D., et al. "A novel process for making alkaline iron oxide nanoparticles by a solvo thermal approach." *Journal of Structural Chemistry* 55.3(2014): 525–529.
38. Hui, Fei, and Mario Lanza. "Scanning probe microscopy for advanced nanoelectronics." *Nature Electronics* 2.6(2019): 221–229.
39. Collomb, David, et al. "Nanoscale CVD graphene hall probes for high resolution scanning probe microscopy." *APS* 2019(2019): L23-002.
40. Teo, Yik R., Yuen Yong, and Andrew J. Fleming. "A comparison of scanning methods and the vertical control implications for scanning probe microscopy." *Asian Journal of Control* 20.4(2018): 1352–1366.
41. Mitta, Saisameera, and Ulrich Stimming. "Surface characterization of commercial Li-Ion battery electrodes using scanning probe microscopy." *Electrochemical Conference on Energy and the Environment (ECEE 2019): Bioelectrochemistry and Energy Storage* (July 21–26, 2019). ECS, 2019.
42. Baykara, Mehmet Z., et al. "Low-temperature scanning probe microscopy." *Springer Handbook of Nanotechnology.* Springer, 2017, 769–808.
43. Plank, Harald, et al. "Focused electron beam-based 3D nanoprinting for scanning probe microscopy: a review." *Micromachines* 11.1(2020): 48.
44. Angeloni, Livia, et al. "Identification of nanoparticles and nanosystems in biological matrices with scanning probe microscopy." *Wiley Interdisciplinary Reviews: Nanomedicine and Nanobiotechnology* 10.6(2018): e1521.
45. Shur, V. Ya. "Study of ferroelectric domains by scanning probe microscopy." *Scanning Probe Microscopy. Russia-China Workshop on Dielectric and Ferroelectric Materials—Ekaterinburg,* 2019. Ural Federal University, 2019.
46. Lambert, Joseph B., Eugene P. Mazzola, and Clark D. Ridge. *Nuclear Magnetic Resonance Spectroscopy: An Introduction to Principles, Applications, and Experimental Methods.* John Wiley & Sons, 2019.

47. Larsen, Michael S., et al. "Nuclear magnetic resonance probe system." U.S. Patent No. 9,970,999. 15 May 2018.
48. Khaneja, Navin. "Nuclear Magnetic Resonance." (2020). DOI: 10.5772/intechopen.74899.
49. Ruan, R. Roger, and Paul L. Chen. "Nuclear magnetic resonance techniques." *Bread Staling*. CRC Press, 2018, 113–128.
50. Cote, Rene, and Jean-Michel Parent. "Nuclear magnetic resonance line shapes of electron crystals in 13 C graphene." *APS* 2018 (2018): B35-007.

Section II

Synthesis

6 Synthesis and Fabrication of Graphene/Ag-Infused Polymer Nanocomposite

Chinedu Okoro
Tuskegee University

Zaheeruddin Mohammed
Tuskegee University

Lin Zhang
Auburn University

Shaik Jeelani
Tuskegee University

Zhongyang Cheng
Auburn University

Vijaya Rangari
Tuskegee University

CONTENTS

6.1	Introduction	88
6.2	Materials and Methods	90
	6.2.1 Materials	90
	6.2.2 Graphene Synthesis Using Organoclay as Precursor and Autogenic Pressure Reaction Technique	90
	6.2.3 Graphene–Ag Nanoparticle Synthesis	90
	6.2.4 Polymer Nanocomposite Synthesis	90
6.3	Characterization	91
	6.3.1 Thermogravimetric Analysis (TGA) of Nanoclay	91
	6.3.2 X-Ray Diffraction (XRD) of SG and SG-Ag Nanoparticles	91
	6.3.3 Transmission Electron Microscopy (TEM) of Nanoparticles	91
	6.3.4 X-Ray Photoelectron Spectroscopy (XPS) of SG and SG-Ag Nanoparticles	91
	6.3.5 Raman Spectroscopy of SG and SG-Ag Nanoparticles	91

        6.3.6   Electrical Conductivity of SG and SG-Ag Nanoparticles 91
        6.3.7   Thermal Conductivity of SG and SG-Ag Nanoparticles 92
6.4   Results and Discussion 92
        6.4.1   Thermogravimetric Analysis of Nanoclay 92
        6.4.2   X-Ray Diffraction (XRD) Analysis of SG and SG-Ag Nanoparticles 94
        6.4.3   Transmission Electron Microscopy (TEM) and Energy-
                  Dispersive Spectroscopy (EDS) of Graphene–Ag Nanoparticles 94
        6.4.4   X-Ray Photoelectron Spectroscopy (XPS) of SG and SG-Ag
                  Nanoparticles 95
        6.4.5   Raman Spectroscopy of SG and SG-Ag Nanoparticles 95
        6.4.6   Electrical Conductivity of SG and SG-Ag Nanoparticles 96
        6.4.7   Thermal Conductivity of SG and SG-Ag Nanoparticles.................... 98
        6.4.8   Thermo-Mechanical Analysis (TMA) of Graphene Nanoparticle
                  Polymer Nanocomposite Systems.. 99
        6.4.9   Thermogravimetric Analysis (TGA) of Graphene Nanoparticle
                  Polymer Composite Systems 100
        6.4.10 Dynamic Mechanical Analysis (DMA) of Graphene
                  Nanoparticle Composite Systems 101
        6.4.11 Flexure 3-Point Bending Analysis... 103
6.5   Summary 104
Acknowledgments.. 105
References.. 105

6.1 INTRODUCTION

Fabrication of nanocomposites incorporating different kinds of nanoscale filler materials into polymer-based matrices has opened up a new area of research. Using nanoscale materials, composites with multifunctional properties and better performance can be developed without making major changes to the manufacturing process [1]. One such nanomaterial that gained extensive attention from researchers all over the world is graphene. The primary reason for such interest is the multi functionality of its 2D atomic crystal that renders its unique properties such as thermal conductivity of around 5,000 W/mK [2], high electron mobility of 250,000 cm^2/Vs at room temperature [3], exceptionally large surface area of 2,630 m^2/g [4], and high modulus of elasticity of about 1 TPa along with good electrical conductivity [5]. Graphene, which is also considered as the mother of all graphitic forms, is actually a single layer of carbon atoms that are held together by overlapping sp^2 hybrid bonds [6]. The remarkable properties of graphene come from the 2P orbitals, which form the π bonds that delocalize over the sheet of carbon in the graphene. This leads to its superior properties that are discussed earlier [7].

There are many routes to synthesize graphene; the simplest of all of them is through mechanical exfoliation of graphite. This method was first used to isolate graphene that led to the Nobel Prize in 2010. The quality of graphene made using this method is considered the highest. However, this method is useful only for small-scale lab experiments, as yield is very low and scaling up is very difficult [8]. Chemical vapor deposition (CVD) [9], liquid-phase exfoliation [10, 11], electrochemical exfoliation [12, 13], and chemical reduction [14–17] are few other ways to synthesize graphene [18]. Autogenic pressure reaction (APR) is an approach where carbon-containing precursors are carbonized at a very high temperature and pressure [19]. Here, the thermal disassociation of the hydrocarbons and particle size are controlled by the internal pressure created by the precursor partial pressure. Pol et al. synthesized spherical diamond-like hard carbon from pyrolysis of polyethylene at a high pressure and temperature [20]. Using a similar process, graphitic carbon was synthesized from starch-based packaging waste material and silica/carbon-rich rice husks [21, 22]. The synthesis of graphene using the APR process requires a highly layered precursor material that is rich in carbon.

Epoxy resins are generally considered a good matrix material due to their good tensile modulus, glass transition temperature, and thermal stability. However, they lack good strain to failure and toughness [23]. To improve the toughness and ductility of epoxy, one method is to disperse toughening in cured resins that consist of epoxy monomers and hardener [24, 25]. While developing composites using nanoscale graphene, the general idea is to use high-modulus graphene and low-modulus matrix material to develop nanocomposites with significant reinforcement ability. The outcome of such materials is improved mechanical, electrical, and thermal properties. It was found that at 4 wt. % loading, graphene nanoplatelets were able to improve fracture toughness of epoxy matrix by 85% [26]. In another study with a loading as low as 0.5 wt. %, the storage modulus of epoxy nanocomposites increased 50% when compared to neat epoxy samples [27]. A study reported that with a loading of only 0.4 wt. % imidazole-functionalized graphene reinforced with epoxy matrix, tensile strength and modulus were enhanced by 97% and 12%, respectively [28].

Sometimes reinforcing the nanocomposite with one material is not enough to perform its desired functionality; in such circumstances, hybrid fillers are used. The primary goal for developing a hybrid material is to take advantage of favorable intrinsic properties of individual materials and achieve an overall synergistic effect. Yang et al. developed nanocomposites reinforced with multi walled carbon nanotubes (MWCNT) and multi-graphene platelets (MGP) with a loading of 0.1 and 0.9 wt.%, respectively, in epoxy matrix. They have found that the mechanical properties such as tensile strength, modulus, and elongation increased. Thermal conductivity of the nanocomposites also increased. The primary reason for the increase in properties was attributed to better dispersion of graphene due to the presence of nanotubes on their surface leading to the formation of a 3D structure [29]. In another study, it was reported that hybrid loading of SiO_2/GO increased the tensile strength of epoxy composite up to 31% at a loading of only 1.5 wt. % [30]. Graphene and nanoclay were used to improve mechanical and thermal properties of epoxy composites. There was a 29% increase in storage modulus for binary reinforcement at 0.1 wt.% GNP and 3 wt.% MMT loading [31]. Electrically conductive adhesives were developed using Ag nanowires, nanoparticles, and graphene platelets in epoxy matrix. It was found that adhesives loaded with 0.8 wt.% had the lowest resistivity of 3.01×10^{-4} Ω cm [32].

In the current work, hybrid nanoparticles of graphene/silver are developed using low-cost precursor materials. Thus, synthesized nanoparticles are used to make nanocomposites of epoxy-based resin system. Nanocomposites were then tested for electrical, thermal, mechanical, and viscoelastic properties.

6.2 MATERIALS AND METHODS

6.2.1 Materials

Graphene precursor Cloisite 20A was purchased from Southern Clay Products, Inc., TX. Grade C1 commercial graphene was purchased from Graphene Supermarket, NY. DGEBA-based epoxy resin SC-15 was purchased from Applied Poleramic Inc., CA. Plasticizer EP9009 was from Eager Plastics, and dimethylformamide (DMF), silver acetate, and copper acetate were all purchased from Sigma-Aldrich, MO, USA.

6.2.2 Graphene Synthesis Using Organoclay as Precursor and Autogenic Pressure Reaction Technique

Cloisite 20A was added along with 1 wt.% catalyst (copper acetate) by weight into a capped Swagelok. The Swagelok was then placed in a tube furnace and heated to 1,000 °C, and then left to dwell for 1 h. The black powder was removed, and particles were ground using a mortar and pestle. The particles were washed with nitric acid by magnetic stirring for 24 h to remove metallic impurities. Particles were then washed with water several times and finally centrifuged at 10,000 rpm for 10 min in ethanol. Particles were dried under vacuum for 24 h.

6.2.3 Graphene–Ag Nanoparticle Synthesis

Graphene obtained from autogenic pressure reaction was mixed with 1 wt.% silver acetate in DMF solution. The solution was sonicated for 10 min and then transferred to CEM microwave for silver acetate reduction and deposition onto graphene nanoplatelets. The CEM is programmed to run at 150 W and 50 psi. Particles are then removed from the CEM and centrifuged in ethanol at 10,000 rpm for three 10-min cycles. Silver-decorated graphene (SG-Ag) is then placed in a desiccator for drying.

6.2.4 Polymer Nanocomposite Synthesis

SC-15A was measured and mixed with plasticizer EP9009 (10:1). This mixture was mechanically mixed for 5 min. SC-15B was measured according to stoichiometric ratio (10:3) and mixed with the desired percentage of SG-Ag nanoparticles [33–36]. The nanoparticles were dispersed in solution for 20 min using ultrasonic irradiation with a 0.25-inch probe. Both mixtures were then combined and mixed using mechanical and magnetic stirring for 20 min. Final mixture was then placed in a vacuum oven for degasification for 30–45 min, respectively. The mixture was then poured into the desired mold for curing. The mixture was cured at 60 °C for 2 h and then post-cured at 100 °C for 2 h. Samples were then removed from the mold and cut precisely according to ASTM standard.

6.3 CHARACTERIZATION

6.3.1 THERMOGRAVIMETRIC ANALYSIS (TGA) OF NANOCLAY

Thermogravimetric analysis (TGA) is commonly used to determine the decomposition temperatures, residual components, absorbed moisture content, and the amount of inorganic filler in polymer or composite material composition. TGA was performed using Mettler Toledo TGA/SDTA851 operating in nitrogen gas at a heat rate of 10 °C /min, from ambient to 1,000 °C. The TGA samples were in powder form, and the weights were between 5 and 10 mg. TGA data were analyzed using the STAR^e Evaluation software.

6.3.2 X-RAY DIFFRACTION (XRD) OF SG AND SG-Ag NANOPARTICLES

X-ray diffraction was performed using Rigaku RINT2100 X-ray diffractometer with monochromatic CuK α radiation ($\lambda = 0.154056$ nm) generated at 40 kV and 30 mA. Scan parameters were as follows: scanning range, 3°–80°; sampling width, 0.02; and scan speed, 5°/min. Analysis was performed using Jade 9.0 software.

6.3.3 TRANSMISSION ELECTRON MICROSCOPY (TEM) OF NANOPARTICLES

JEOL-JEM 2010 transmission electron microscope with Oxford INCA 100 energy-dispersive spectroscopy (EDS) was used to characterize the nanoparticles.

6.3.4 X-RAY PHOTOELECTRON SPECTROSCOPY (XPS) OF SG AND SG-Ag NANOPARTICLES

XPS was performed using a Kratos AXIS 165 multitechnique electron spectrometer. A 15-keV electron gun was used for AES analysis and Auger or secondary electron imaging. Instrument control, data acquisition, and data analysis were performed through the Kratos Vision 1.5 software.

6.3.5 RAMAN SPECTROSCOPY OF SG AND SG-Ag NANOPARTICLES

Raman spectroscopy was performed using the DXR Raman microscope. The laser was 532 nm with a 532-nm full-range grating. Laser power at the sample was 1 mW with 100× microscope objective and 50-mm pinhole slit.

6.3.6 ELECTRICAL CONDUCTIVITY OF SG AND SG-Ag NANOPARTICLES

The Agilent 4294A impedance analyzer was employed to measure the dielectric property of the samples over a frequency range of 100–1 MHz using the Cp~D function. For the characterization of the dielectric properties for the sample, the samples were sputtered with gold on the top and bottom surfaces as electrodes using a Pelco

SC-6 sputter coater. To obtain a uniform coating of gold for the electrode, four times of 30-s coating of each side was necessary, which resulted in a gold layer with a thickness about 40–60 nm.

6.3.7 Thermal Conductivity of SG and SG-Ag Nanoparticles

The thermal conductivity measurements were performed using the C-Therm TCi. The composite and ceramic template was used. Effusivity is defined as the square root of the product of thermal conductivity, k, density, ρ, and specific heat capacity, Cp (i.e., $\sqrt{k \cdot \rho \cdot c_p}$), and has units of $\dfrac{W \sqrt{s}}{m^2 K}$. The effusivity measurements are taken directly, and the conductivity is calculated from this measurement. Contact medium was distilled water. Three scans were performed for each sample.

6.4 RESULTS AND DISCUSSION

6.4.1 Thermogravimetric Analysis of Nanoclay

The thermogravimetric analysis (TGA) illustrates the nanoclay transformation and weight loss. It is seen early in decomposition that there are three distinct decomposition regions before 1,000 °C. These are attributed to the loss of organic components in the nanoclay (Figure 6.1).

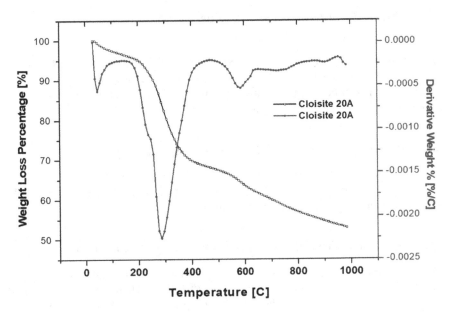

FIGURE 6.1 TGA weight loss and derivative curve of nanoclay.

Synthesis and Fabrication of Graphene/Ag

From Figure 6.2, the amount of modifier or carbon content expected is predicted once all organic components have decomposed.

In order to understand the carbon content, the modifier concentration was used and HT carbon amount was quantified:

$$C18 = 12\frac{g}{atom} * 18 \text{ atom} = 216g * 0.65 = 140.4g$$

$$C16 = 12\frac{g}{atom} * 16 \text{ atom} = 192g * 0.3 = 57.6g$$

$$C14 = 12\frac{g}{atom} * 14 \text{ atom} = 168g * 0.05 = 8.4g$$

Therefore,

$$1 \text{ mol of HT} = 206.4g$$

Hence, the modifier molar mass can be calculated:

$$1 \text{ mol of } (CH_3)_2 N(HT)_2 = 30g + 14g + 412.8g = 456.8g$$

$$0.095 \text{ eq} * 456.8g = 43.4g$$

From theoretical calculation for every 100 g of clay, there is ~43 g of modifier present.

Therefore, there will be ~45% carbon seen solely from the modifier component. However, the TGA curve indicates a slightly larger percentage of mass remaining. This is due to the remaining metallic components still being present. Hence, additional processing was performed to remove the metallic impurities.

Treatment/Properties:	Organic Modifier (1)	Modifier Concentration	% Moisture	% Weight Loss on Ignition
Cloisite® 20A	2M2HT	95 meq/100g clay	< 2%	38%

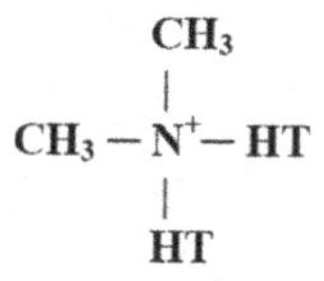

$$CH_3 - \underset{\underset{HT}{|}}{\overset{\overset{CH_3}{|}}{N^+}} - HT$$

Where HT is Hydrogenated Tallow (~65% C18; ~30% C16; ~5% C14)

FIGURE 6.2 Block structure of modifier within Cloisite 20A.

6.4.2 X-Ray Diffraction (XRD) Analysis of SG and SG-Ag Nanoparticles

To assess the composition and crystallinity of the graphene derived from nanoclay, XRD analysis was performed. The SG-Ag particles were also analyzed to confirm the reduction of silver acetate to silver within the SG-Ag particle system. The synthesized graphene profile matches the major peaks of the graphite JCPDS profile. SG-Ag nanoparticles show equivalent matching peaks for silver as well as some less prominent graphite peaks (Figure 6.3).

There is a presence of SiO_2 which is depicted by the prominent peak seen in the synthesized graphene at 22° [2θ] mark. This is expected based on the chemical structure of the precursor nanoclay with silica being an inherently present constituent component. This compound can be removed with additional processing but is not needed for the desired application of this study.

6.4.3 Transmission Electron Microscopy (TEM) and Energy-Dispersive Spectroscopy (EDS) of Graphene–Ag Nanoparticles

TEM images show that the graphene sheets have been fabricated, and the micrographs confirm the presence of Ag decorated on the surface or the graphene platelets (Figure 6.4). Micrographs depict multiple sheets along with spherical and hexagonal silver particles deposited on the surface (Figure 6.4b). For nanostructure comparison, a TEM image was performed on a commercially available graphene nanosheet (Figure 6.4c).

The micrograph indicates difference in morphology. The commercial graphene seems to be in smaller platelet form, whereas the synthesized graphene sheets appear

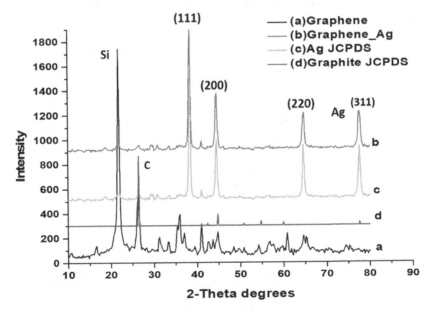

FIGURE 6.3 XRD plot of (a) SG, (b) SG-Ag, (c) Ag standard, and (d) graphite standard.

Synthesis and Fabrication of Graphene/Ag

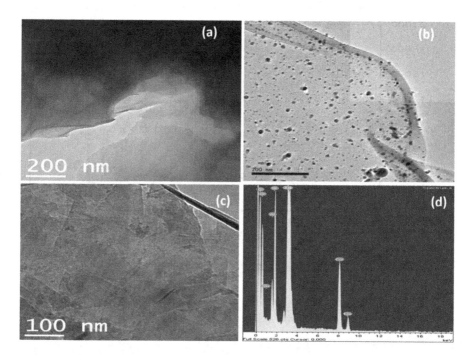

FIGURE 6.4 TEM micrographs of (a) SG nanoparticles, (b) SG-Ag nanoparticles, (c) commercial graphene nanoparticles, and (d) EDS spectrum of SG-Ag nanoparticles.

to have larger surface area between grain boundaries. The energy-dispersive spectroscopy (EDS) of SG-Ag nanoparticles confirms the presence of carbon, silver, and oxygen along with some silicon dioxide (Figure 6.4d). The copper peaks are a result of the copper grid that was used for analysis [37].

6.4.4 X-Ray Photoelectron Spectroscopy (XPS) of SG and SG-Ag Nanoparticles

XPS confirms that Ag at zero oxidation state has successfully bonded to the graphene nanoparticles. Zero-oxidation state silver formation is confirmed by the resonance peaks seen at the 4p (25 eV), 3d (370–385 eV), and 3p (580–610 eV) orbitals [37].

The carbon peak is present in both samples at ~285 eV. Si is present as indicated earlier due to the chemical morphology of the precursor material. In addition, there seems to be an oxygen band present, seen in Figure 6.5b. This may be attributed to oxidation taking place during the organic disassociation forming oxide regions along the grain boundaries of the graphene sheet.

6.4.5 Raman Spectroscopy of SG and SG-Ag Nanoparticles

The Raman spectrum indicates an oxidized form of graphene. Graphene oxide is produced through an oxygen-producing chemical reaction within the layers of a graphite

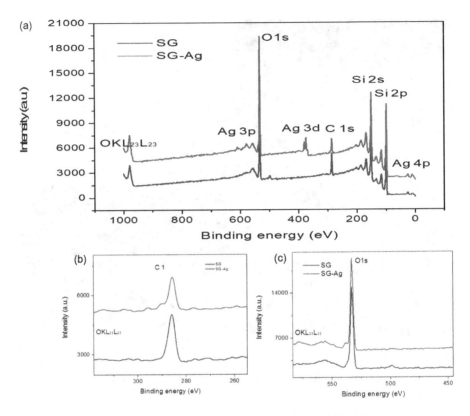

FIGURE 6.5 (a) XPS spectrum of SG and SG-Ag nanoparticles, (b) O1s band presence, and (c) carbon excitation band in neat and SG-Ag nanoparticles.

crystal. The nanoclay structure, size, and morphology promoted small crystal formation of oxidized graphene. Raman spectra of the materials (Figure 6.6) show strong D (1,600 cm^{-1}) and G (1,350 cm^{-1}) peaks, suggesting very small crystal sizes [38].

These particles can be further reduced for additional applications; however, it is theorized that these particles produce electrical and thermal properties similar to those of high-quality graphene. Additionally, the presence of silver seems to increase the intensities on the graphene spectrum.

6.4.6 Electrical Conductivity of SG and SG-Ag Nanoparticles

Dielectric analysis was performed to assess the dielectric constant of the SG and SG-Ag nanoparticles. Figure 6.7 depicts the dielectric constant and loss constant of (a) 1% SG-Ag, (b) 1% SG, (c) 2% SG-Ag, (d) 2% SG, (e) 5% SG-Ag, (f) 5% SG, and (g) neat SC-15 polymer nanocomposites. From the data, the dielectric constant increases as the nanoparticle content is increased in both SG (b, d, f) and SG-Ag (a, c, e) samples. There was a significant increase in the dielectric constant seen in the 5% SG (f) samples (130.96) of 398% when compared to the neat SC-15 (26.28) (g) samples. This enhancement may be attributed to the particle surface area and the

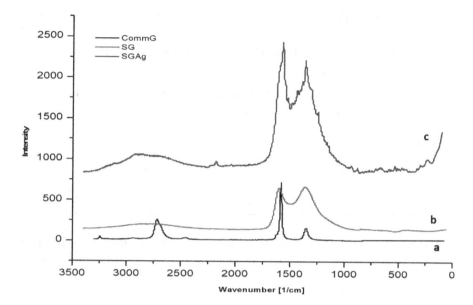

FIGURE 6.6 Raman spectrum of (a) commercial graphene, (b) synthesized graphene, and (c) synthesized graphene/silver nanoparticles.

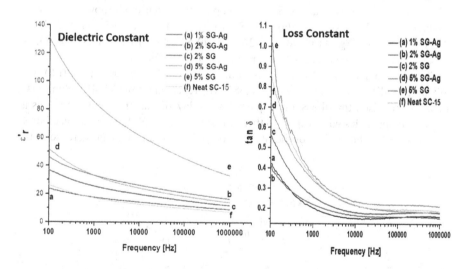

FIGURE 6.7 Dielectric analysis and loss constant of (a) 1% SG-Ag, (b) 1% SG, (c) 2% SG-Ag, (d) 2% SG, (e) 5% SG-Ag, (f) 5% SG, and (g) neat SC-15.

reduction that may have taken place due to the secondary processing and addition of the SG-Ag nanoparticles (Table 6.1).

In addition, the percolation threshold may have been reduced in the SG-Ag particles due to additional processing, whereas the SG nanoparticles were kept intact, thereby being able to perform better as conductors.

TABLE 6.1
Dielectric Analysis of SG and SG-Ag Polymer Nanocomposites

	Dielectric Constant		Loss Constant	
Sample	ε'_r	Standard Deviation	Tan δ	Standard Deviation
Neat SC-15	26.28	1.64	0.86	0.04
1% SG-Ag	23.05	2.28	0.41	0.07
2% SG-Ag	46.09	1.40	0.42	0.01
5% SG-Ag	51.53	0.89	0.69	0.03
1% SG	35.57	1.27	0.47	0.00
2% SG	36.53	6.84	0.56	0.04
5% SG	130.96	9.35	1.02	0.02

The loss constant depicts the standard loss curve anticipated for the supporting samples. There is no significant loss seen in any samples indicating accurate dielectric results.

6.4.7 Thermal Conductivity of SG and SG-Ag Nanoparticles

The thermal conductivity tests that were performed give a direct comparison of the SG versus the CG as well as the SG-Ag nanoparticles. Figure 6.8 shows the thermal conductivity and effusivity of (a) neat SC-15, (b) 10% EP, (c) 1% SG-Ag, (d) 2% SG-Ag, (e) 5% SG-Ag, (f) 1% SG, (g) 2% SG, (h) 5% SG, (i) 1% CG, (j) 2% SG, (k) 5% CG, and (l) 50% SG- Ag polymer nanocomposites (Table 6.2).

The SG nanoparticles (f-h) are comparable with the CG nanoparticles (i-k) with only a ~10% difference in the 5% SG (0.381 W/mK) (h) and 5% CG (419 W/mK) (k). It can be surmised that the SG nanoparticles display similar transient properties as the CG nanoparticles with respect to thermal conductivity. With the addition of silver nanoparticles to the surface of the SG nanoparticles, there is also an increase in the

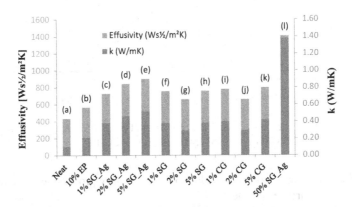

FIGURE 6.8 Thermal conductivity of (a) neat SC-15, (b) 10% EP, (c) 1% SG-Ag, (d) 2% SG-Ag, (e) 5% SG-Ag, (f) 1% SG, (g) 2% SG, (h) 5% SG, (i) 1% CG, (j) 2% SG, (k) 5% CG, and (l) 50% SG-Ag polymer nanocomposites.

TABLE 6.2
Thermal Conductivity of Polymer Nanocomposites

Sample	Effusivity (Ws$^{1/2}$/m²K)	Standard Deviation	k (W/mK)	Standard Deviation
Neat SC-15	435.83	1.43	0.101	0.001
10% EP	566.49	1.56	0.210	0.001
1% SG-Ag	726.48	24.77	0.382	0.023
2% SG-Ag	842.09	25.38	0.461	0.025
5% SG-Ag	899.74	24.34	0.519	0.024
1% SG	753.91	38.24	0.377	0.036
2% SG	658.43	2.19	0.290	0.002
5% SG	758.34	2.97	0.381	0.003
1% CG	779.71	19.52	0.402	0.018
2% CG	658.89	43.67	0.291	0.039
5% CG	797.74	28.20	0.419	0.027

conductivity of the nanoparticle system 5% SG-Ag (e) (conductivity, 0.519 W/mK; effusivity, 900 Ws$^{1/2}$/m²K) up to ~36%, respectively.

6.4.8 THERMO-MECHANICAL ANALYSIS (TMA) OF GRAPHENE NANOPARTICLE POLYMER NANOCOMPOSITE SYSTEMS

Thermo-mechanical analysis was performed to understand the effects of the nanoparticulate systems on the coefficient of thermal expansion of the modified polymer system. Figure 6.9 shows the TMA curves of (a) neat SC-15, (b) 10% EP, (c) 1% SG-Ag, (d) 2% SG-Ag, (e) 5% SG-Ag, (f) 1% SG, (g) 2% SG, (h) 5% SG, (i) 1% CG, (j) 2% SG, (k) 5% CG, and (l) 50% SG-Ag nanocomposites. From the test data, the ratio of dimensional change over temperature change is plotted. Using the analysis software, we are able to take the slope both before and after alpha relaxation and glass transition.

From the test data, the ratio of dimensional change over temperature change is plotted. The coefficient of thermal expansion (CTE) is then calculated using the following formula :

$$\alpha = \frac{1}{L_o} * \frac{\Delta L}{\Delta T}$$

where the slope of the initial portion of the curve gives the value for dL/dT and L is the thickness of the samples.

Thermo-mechanical analysis depicts a standard trend and slight reduction in the coefficient of thermal expansion (CTE) throughout the nanophased samples as nanoparticle content was increased.

The reduction in CTE values can be attributed to the thermal conductivity of the graphite nanoparticles. However, of all the systems, 5% SG (h), 5% CG (k), and 5% SG-Ag (e) showed the greatest reduction in CTE values prior to glass transition at (0.038 1/°C) 45%, (0.059 1/°C) 15%, and (0.048 1/°C) 31%, respectively (Table 6.3).

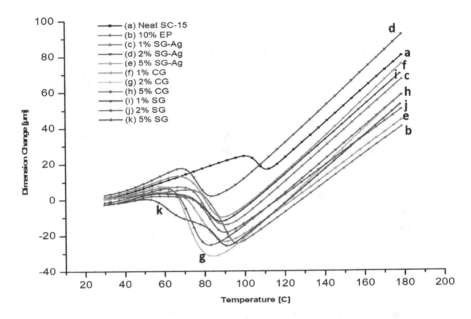

FIGURE 6.9 TMA curves of neat and graphene nanophased polymer composites.

TABLE 6.3
Thermo-Mechanical Properties

Sample	CTE Pre-t_g (1/°C)	Standard Deviation	CTE Post-t_g (1/°C)	Standard Deviation
Neat SC-15	0.060	0.035	0.195	0.012
10% EP9009	0.070	0.014	0.192	0.004
1% SG-Ag	0.047	0.007	0.189	0.002
2% SG-Ag	0.070	0.031	0.186	0.008
5% SG-Ag	0.048	0.015	0.180	0.006
1% CG	0.069	0.014	0.180	0.012
2% CG	0.065	0.011	0.194	0.008
5% CG	0.059	0.005	0.184	0.003
1% SG	0.068	0.014	0.187	0.002
2% SG	0.069	0.009	0.190	0.004
5% SG	0.038	0.004	0.188	0.010

6.4.9 THERMOGRAVIMETRIC ANALYSIS (TGA) OF GRAPHENE NANOPARTICLE POLYMER COMPOSITE SYSTEMS

Thermogravimetric analysis was performed for (a) neat SC-15, (b) 10% EP, (c) 1% SG-Ag, (d) 2% SG-Ag, (e) 5% SG-Ag, (f) 1% SG, (g) 2% SG, (h) 5% SG, (i) 1% CG, (j) 2% SG, and (k) 5% CG to understand the effects of the nanoparticulates

Synthesis and Fabrication of Graphene/Ag

TABLE 6.4
Thermogravimetric Properties

Sample	t_5	Standard Deviation	t_{50}	Standard Deviation	$\%_r$
Neat SC-15	304.00	5.66	365.00	0.59	5.00
10% EP9009	270.00	7.07	338.00	0.21	8.00
1% SG-Ag	281.59	10.17	361.49	0.62	4.65
2% SG-Ag	298.18	2.58	361.09	0.42	3.81
5% SG-Ag	279.33	8.78	366.90	1.34	6.83
1% CG	254.09	3.52	360.59	0.53	4.91
2% CG	248.63	6.84	358.29	1.41	4.85
5% CG	258.86	0.45	358.85	0.41	7.34
1% SG	301.81	3.75	361.98	1.15	4.43
2% SG	297.69	0.71	359.97	0.11	7.27
5% SG	306.74	0.51	362.32	0.60	4.88

on the decomposition temperature of the modified polymer system. Data analysis from the weight loss curve of the thermogravimetric analysis indicates there is an increase in the degradation onset point in both the SG (f–h) and SG-Ag (c–e) nanophased samples. Largest improvements were seen in 5% SG (h) (307°C; ~13%) and 2% SG-Ag (d) (298°C; 10%, respectively. This may be attributed to the larger surface area seen in these two systems, hence improved dispersion within the polymer system. Subsequently, there is an increase in overall decomposition temperature seen in all nanophased systems with the most significant being the 5% SG-Ag (367°C) (e) at ~8%. The increase in decomposition temperature can be attributed to the thermally conductive nature of the particulate systems (Table 6.4).

The ability to conduct heat allows the particles to perform as a throughput carrier for heat, thus increasing the heat capacity of the polymer system producing a slight delay in polymer chain movement.

The derivative weight curves indicate a reduction in thermal stability as the plasticizer is introduced into the system; however, there is continued stability as the nanoparticles were added to the plasticized system.

6.4.10 DYNAMIC MECHANICAL ANALYSIS (DMA) OF GRAPHENE NANOPARTICLE COMPOSITE SYSTEMS

Dynamic mechanical analysis (DMA) was performed on all neat and nanophased plasticized samples. Figure 6.10 displays the DMA storage modulus curves of (a) neat SC-15, (b) 10% EP, (c) 1% SG-Ag, (d) 2% SG-Ag, (e) 5% SG-Ag, (f) 1% SG, (g) 2% SG, (h) 5% SG, (i) 1% CG, (j) 2% SG, and (k) 5% CG polymer nanocomposites. Data interpretation indicates that there is a fairly linear increase in storage modulus throughout all systems. The 5% SG-Ag (e) (2,788 MPa) showed the highest overall storage modulus improvement at ~19%, respectively (Table 6.5).

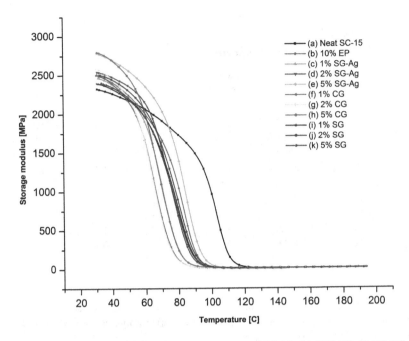

FIGURE 6.10 DMA storage modulus curves of (a) neat SC-15, (b) 10% EP, (c) 1% SG-Ag, (d) 2% SG-Ag, (e) 5% SG-Ag, (f) 1% SG, (g) 2% SG, (h) 5% SG, (i) 1% CG, (j) 2% SG, and (k) 5% CG.

TABLE 6.5
Dynamic Mechanical Analysis

Sample	Storage Modulus (MPa)	Standard Deviation	Loss Modulus (MPa)	Standard Deviation	Tan δ	Standard Deviation	t_g (°C)	Standard Deviation
Neat SC-15	2324.80	144.14	234.13	10.97	0.82	0.00	114.70	1.31
10% EP9009	2500.21	43.22	262.78	6.69	0.94	0.02	95.00	0.85
1% SG-Ag	2463.50	30.41	229.20	2.26	0.81	0.01	87.15	0.21
2% SG-Ag	2538.50	71.42	241.00	5.52	0.82	0.00	91.61	0.33
5% SG-Ag	2788.33	181.75	272.23	8.02	0.88	0.04	98.99	1.42
1% CG	2506.50	136.47	247.35	7.28	0.80	0.01	82.10	0.39
2% CG	2581.50	43.13	263.30	4.53	0.79	0.00	81.08	0.08
5% CG	2794.00	318.20	279.15	30.48	0.83	0.00	85.42	0.95
1% SG	2394.00	89.10	222.10	0.42	0.84	0.03	94.67	1.34
2% SG	2501.00	96.17	231.95	9.26	0.80	0.01	94.05	1.94
5% SG	2401.50	91.22	224.75	12.52	0.79	0.02	92.97	0.36

The cross-link density was calculated using the rubber plateau region of the storage modulus curve [39,40]:

$$E_r = 3RT_r v_t$$

Synthesis and Fabrication of Graphene/Ag

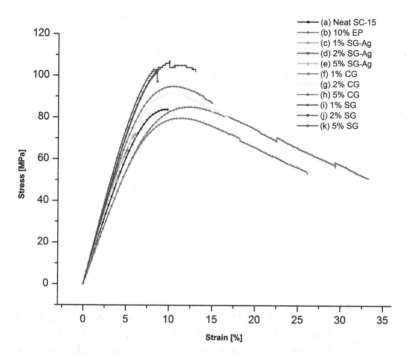

FIGURE 6.11 Stress–strain curves of (a) neat SC-15, (b) 10% EP, (c) 1% SG-Ag, (d) 2% SG-Ag, (e) 5% SG-Ag, (f) 1% SG, (g) 2% SG, (h) 5% SG, (i) 1% CG, (j) 2% SG, and (k) 5% CG polymer nanocomposites.

where E_r = storage modulus (MPa); R = Avogadro's number (m^3Pa·K/mol); T_r = temperature (°K); and v_t = cross-link density (mol·m^3).

Therefore, $E_t = \dfrac{v_r}{3RT_r}$.

Cross-link density analysis shows 5% CG (h) samples have the highest cross-linkage (1,458 mol/m^3) within the polymer system. These values directly coincide with the storage modulus values. Oddly, the 5% SG-Ag (e) samples have relatively similar storage modulus values; however, the cross-link density (704 mol/m^3) is significantly lower. The CG nanoparticles have a significantly lower density when compared to the SG nanoparticles. The higher density SG is due in part to the remaining SiO_2. Therefore, the cross-link density results can be somewhat misleading due to the increased rubbery plateau, which subsequently increases the cross-link density due to the proportionality. It can then be surmised that the reinforcement capabilities of the SG and SG-Ag nanoparticulate systems exceed those of the CG nanoparticles. This is supported by the flexure results seen in Figure 6.11.

6.4.11 Flexure 3-Point Bending Analysis

Stress–strain analysis gives a clear understanding of the mechanical properties of the polymer nanocomposites under static load. Figure 6.11 shows the stress–strain

TABLE 6.6
Cross-Link Density of Polymer Nanocomposites

Sample	Rubbery Plateau Modulus E_r (MPa)	Temperature T_r (K)	Cross-Link Density ν_t (mol/m³)
Neat SC-15	8.25	418.15	791.03
10% EP9009	3.31	418.15	316.89
1% SG-Ag	9.64	418.15	924.05
2% SG-Ag	11.55	418.15	1107.24
5% SG-Ag	7.35	418.15	704.95
1% CG	10.08	418.15	966.85
2% CG	9.82	418.15	942.00
5% CG	15.22	418.15	1458.90
1% SG	12.90	418.15	1236.58
2% SG	12.56	418.15	1203.90
5% SG	14.28	418.15	1369.15

curves of (a) the neat SC-15, (b) 10% EP, (c) 1% SG-Ag, (d) 2% SG-Ag, (e) 5% SG-Ag, (f) 1% SG, (g) 2% SG, (h) 5% SG, (i) 1% CG, (j) 2% SG, and (k) 5% CG polymer nanocomposites. Data interpretation depicts SG (f–h) nanoparticles outperforming the other constituent materials in overall load capacity (Table 6.6).

This may be attributed to several factors; one could presume a better interfacial interaction between that of the SG nanoparticles with the matrix when compared to the SG-Ag nanoparticles. Based on the morphological variances within the hybrid particle such as hexagonally, and circularly shaped silver when compared to solely multilayer graphene. Adversely, it can be surmised that the CG nanoparticles had an extremely regressive effect on the polymer system shown by the reduction in stress and strain by up to (54.40 MPa) 57% and (4.24 %) 90%, respectively (Table 6.6).

This can be ascribed to the chemical morphology of the spz freestanding orbitals on the commercial graphene as compared to the synthesized graphene (Table 6.7).

The processed graphene has oxide regions that help create a better cross-linkage with the matrix. Also, the density variance may factor into the particulate amount that would have then led to potentially poor dispersion within the matrix.

6.5 SUMMARY

Graphene nanoparticles were successfully synthesized using a Cloisite 20A nanoclay and a copper catalyst in a capped Swagelok autogenic pressure reaction. A yield of ~43% graphene was observed per 100 g of clay. XRD confirmed the graphene and silver crystalline peaks. TEM images confirmed graphene structure and SG-Ag structure along with particle distribution on the surface compared with commercial graphene. EDS also confirmed the elemental structure of SG-Ag nanoparticles. XPS confirmed the Ag zero oxidation state as well as carbon peak indicating graphene structure. XPS also confirmed silica presence and the first indication to potential oxide regions on the graphene. Raman analysis confirmed oxidized graphene compared with commercial graphene.

TABLE 6.7
Stress–Strain Data of Polymer Nanocomposites

Sample	Stress (MPa)	Standard Deviation	Strain (%)	Standard Deviation	Modulus (GPa)	Standard Deviation
Neat SC-15	83.89	3.50	9.81	2.99	2.65	0.06
10% EP	84.99	2.33	40.49	10.38	2.68	0.16
1% CG	72.50	3.55	6.09	0.46	2.86	0.03
2% CG	65.41	3.51	5.16	0.38	3.05	0.08
5% CG	54.40	4.19	4.24	0.20	3.17	0.23
1% SG-Ag	79.69	2.10	26.07	4.08	4.35	0.12
2% SG-Ag	91.39	2.14	16.96	1.14	5.60	0.23
5% SG-Ag	94.96	0.85	14.95	0.70	6.20	0.46
1% SG	106.66	3.37	12.98	2.06	2.93	0.23
2% SG	102.51	7.69	8.66	0.03	3.06	0.27
5% SG	104.31	0.40	8.93	0.46	3.22	0.07

Polymer nanocomposites were successfully fabricated using the optimized polymer system and commercial and synthesized nanoparticles. Dielectric analysis results showed improvements in all samples, and the highest was seen in the 5% SG samples of 398%. Thermal conductivity measurements show SG nanoparticles to be viable when compared to CG nanoparticles with only ~10% difference. With the addition of silver in the SG-Ag nanoparticle system, the thermal conductivity has increased ~36%. TMA of the polymer nanocomposites indicated a reduction in CTE at 5% SG of 45%. TGA results confirmed an increase in degradation onset point in both SG and SG-Ag samples. Also overall decomposition temperature was improved in the 5% SG-Ag samples of ~8%. DMA results confirmed improvements in the storage modulus, and the highest (19%) was observed in the 5% SG-Ag polymer nanocomposites. Flexure analysis of the polymer nanocomposites indicated that the SG nanoparticles outperformed all other particles. This was ascribed to the interfacial interaction variances of the SG-Ag nanoparticles in comparison. In addition, it was noted the large digression in mechanical strength and strain seen in the CG polymer nanocomposites with reductions up to 57% and 90%, respectively.

ACKNOWLEDGMENTS

The authors would like to acknowledge the financial support of NSF-RISE #1459007, NSF-PREM# 1827690 NSF-CREST#, 1735971, and NSF-MRI-1531934.

REFERENCES

1. Sun, Jing, Chunxiao Wang, Tingting Shen, Hongchen Song, Danqi Li, Rusong Zhao, and Xikui Wang. "Engineering the dimensional interface of BiVO4-2D reduced graphene oxide (RGO) nanocomposite for enhanced visible light photocatalytic performance." Nanomaterials 9, no. 6 (2019): 907.
2. Balandin, Alexander A., Suchismita Ghosh, Wenzhong Bao, Irene Calizo, Desalegne Teweldebrhan, Feng Miao, and Chun Ning Lau. "Superior thermal conductivity of single-layer graphene." Nano Letters 8, no. 3 (2008): 902–907.

3. Novoselov, Kostya S., Andre K. Geim, Sergei Vladimirovich Morozov, Da Jiang, Michail I. Katsnelson, I. V. A. Grigorieva, S. V. B. Dubonos, and A. A. Firsov. "Two-dimensional gas of massless Dirac fermions in graphene." Nature 438, no. 7065 (2005): 197–200.
4. Zhu, Yanwu, Shanthi Murali, Weiwei Cai, Xuesong Li, Ji Won Suk, Jeffrey R. Potts, and Rodney S. Ruoff. "Graphene and graphene oxide: synthesis, properties, and applications." Advanced Materials 22, no. 35 (2010): 3906–3924.
5. Lee, Changgu, Xiaoding Wei, Jeffrey W. Kysar, and James Hone. "Measurement of the elastic properties and intrinsic strength of monolayer graphene." Science 321, no. 5887 (2008): 385–388.
6. Duplock, Elizabeth J., Matthias Scheffler, and Philip JD Lindan. "Hallmark of perfect graphene." Physical Review Letters 92, no. 22 (2004): 225502.
7. Sheehy, Daniel E., and Jörg Schmalian. "Optical transparency of graphene as determined by the fine-structure constant." Physical Review B 80, no. 19 (2009): 193411.
8. Geim, Andre K. "Nobel lecture: random walk to graphene." Reviews of Modern Physics 83, no. 3 (2011): 851.
9. Li, Xuesong, Weiwei Cai, Jinho An, Seyoung Kim, Junghyo Nah, Dongxing Yang, Richard Piner et al. "Large-area synthesis of high-quality and uniform graphene films on copper foils." Science 324, no. 5932 (2009): 1312–1314.
10. Hernandez, Yenny, Valeria Nicolosi, Mustafa Lotya, Fiona M. Blighe, Zhenyu Sun, Sukanta De, I. T. McGovern et al. "High-yield production of graphene by liquid-phase exfoliation of graphite." Nature Nanotechnology 3, no. 9 (2008): 563–568.
11. Niu, Liyong, Jonathan N. Coleman, Hua Zhang, Hyeonsuk Shin, Manish Chhowalla, and Zijian Zheng. "Production of two-dimensional nanomaterials via liquid-based direct exfoliation." Small 12, no. 3 (2016): 272–293.
12. Raccichini, R., A. Varzi, S. Passerini, B. Scrosati. "The role of graphene for electrochemical energy storage." *Nature Materials* 14, no. 3 (2015): 271–279.
13. Abdelkader, A. M., A. J. Cooper, R. A. W. Dryfe, and I. A. Kinloch. "How to get between the sheets: a review of recent works on the electrochemical exfoliation of graphene materials from bulk graphite." Nanoscale 7, no. 16 (2015): 6944–6956.
14. Stankovich, Sasha, Dmitriy A. Dikin, Richard D. Piner, Kevin A. Kohlhaas, Alfred Kleinhammes, Yuanyuan Jia, Yue Wu, SonBinh T. Nguyen, and Rodney S. Ruoff. "Synthesis of graphene-based nanosheets via chemical reduction of exfoliated graphite oxide." Carbon 45, no. 7 (2007): 1558–1565.
15. Wang, Guoxiu, Juan Yang, Jinsoo Park, Xinglong Gou, Bei Wang, Hao Liu, and Jane Yao. "Facile synthesis and characterization of graphene nanosheets." The Journal of Physical Chemistry C 112, no. 22 (2008): 8192–8195.
16. Chen, Wufeng, Lifeng Yan, and P. R. Bangal. "Chemical reduction of graphene oxide to graphene by sulfur-containing compounds." The Journal of Physical Chemistry C 114, no. 47 (2010): 19885–19890.
17. Chua, Chun Kiang, and Martin Pumera. "Chemical reduction of graphene oxide: a synthetic chemistry viewpoint." Chemical Society Reviews 43, no. 1 (2014): 291–312.
18. Papageorgiou, Dimitrios G., Ian A. Kinloch, and Robert J. Young. "Mechanical properties of graphene and graphene-based nanocomposites." Progress in Materials Science 90 (2017): 75–127.
19. Xu, Yangtao, Yameng Zhang, Deyi Zhang, Jiqiang Ma, Wang Yi, Jiwei Zhang, and Hao Shi. "Synthesis of multiwall carbon nanotubes via an inert atmosphere absent autogenetic-pressure method for supercapacitor." Journal of Energy Storage 26 (2019): 100995.
20. Pol, Vilas G., Jianguo Wen, Kah Chun Lau, Samantha Callear, Daniel T. Bowron, Chi-Kai Lin, Sanket A. Deshmukh et al. "Probing the evolution and morphology of hard carbon spheres." Carbon 68 (2014): 104–111.

21. Idrees, Mohanad, Shaik Jeelani, and Vijaya Rangari. "Three-dimensional-printed sustainable biochar-recycled PET composites." ACS Sustainable Chemistry & Engineering 6, no. 11 (2018): 13940–13948.
22. Biswas, Manik C., Shaik Jeelani, and Vijaya Rangari. "Influence of biobased silica/carbon hybrid nanoparticles on thermal and mechanical properties of biodegradable polymer films." Composites Communications 4 (2017): 43–53.
23. Bagheri, R., B. T. Marouf, and R. A. Pearson. "Rubber-toughened epoxies: a critical review." Journal of Macromolecular Science®, Part C: Polymer Reviews 49, no. 3 (2009): 201–225.
24. Mohan, Pragyan. "A critical review: the modification, properties, and applications of epoxy resins." Polymer-Plastics Technology and Engineering 52, no. 2 (2013): 107–125.
25. Jyotishkumar, P., Juergen Pionteck, Ruediger Haessler, Sajeev Martin George, Uroš Cvelbar, and Sabu Thomas. "Studies on stress relaxation and thermomechanical properties of poly (acrylonitrile-butadiene-styrene) modified epoxy–amine systems." *Industrial & Engineering Chemistry Research* 50, no. 8 (2011), 4432–4440.
26. Zaman, Izzuddin, Tam Thanh Phan, Hsu-Chiang Kuan, Qingshi Meng, Ly Truc Bao La, Lee Luong, Osama Youssf, and Jun Ma. "Epoxy/graphene platelets nanocomposites with two levels of interface strength." Polymer 52, no. 7 (2011): 1603–1611.
27. Chen, Zhongxin, and Hongbin Lu. "Constructing sacrificial bonds and hidden lengths for ductile graphene/polyurethane elastomers with improved strength and toughness." Journal of Materials Chemistry 22, no. 25 (2012): 12479–12490.
28. Liu, Wanshuang, Kwang Liang Koh, Jinlin Lu, Liping Yang, Silei Phua, Junhua Kong, Zhong Chen, and Xuehong Lu. "Simultaneous catalyzing and reinforcing effects of imidazole-functionalized graphene in anhydride-cured epoxies." Journal of Materials Chemistry 22, no. 35 (2012): 18395–18402.
29. Yang, S.-Y., W.-N. Lin, Y.-L. Huang, H.-W. Tien, J.-Y. Wang, C.-C. Ma, et al. "Synergetic effects of graphene platelets and carbon nanotubes on the mechanical and thermal properties of epoxy composites." Carbon 49, no. 3 (2011): 793–803.
30. Wang, Rui, Dongxian Zhuo, Zixiang Weng, Lixin Wu, Xiuyan Cheng, Yu Zhou, Jianlei Wang, and Bowen Xuan. "A novel nanosilica/graphene oxide hybrid and its flame retarding epoxy resin with simultaneously improved mechanical, thermal conductivity, and dielectric properties." Journal of Materials Chemistry A 3, no. 18 (2015): 9826–9836.
31. Tcherbi-Narteh, A., Z. Mohammed, M. Hosur and S. Jeelani. "Thermo-mechanical and thermal properties of binary particle nanocomposite exposed to sea water conditioning." *Annals of Materials Science & Engineering* 3, no. 1 (2018): 1029.
32. Wang, Rui, Dongxian Zhuo, Zixiang Weng, Lixin Wu, Xiuyan Cheng, Yu Zhou, Jianlei Wang, and Bowen Xuan. "A novel nanosilica/graphene oxide hybrid and its flame retarding epoxy resin with simultaneously improved mechanical, thermal conductivity, and dielectric properties." Journal of Materials Chemistry A 3, no. 18 (2015): 9826–9836.
33. Jyotishkumar, P., J. Pionteck, L. Häußler, G. Adam, and S. Thomas. "Poly (acrylonitrile-butadiene-styrene) modified epoxy–amine systems analyzed by FTIR and modulated DSC." Journal of Macromolecular Science, Part B 51, no. 7 (2012): 1425–1436.
34. Riaz, Ufana, Arti Vashist, Syed Aziz Ahmad, Sharif Ahmad, and S. M. Ashraf. "Compatibility and biodegradability studies of linseed oil epoxy and PVC blends." Biomass and Bioenergy 34, no. 3 (2010): 396–401.
35. E. Polymers. *EP9009 Material Data Safety Sheet*, Eager Polymers, Chicago, 2012.
36. Sharmin, Eram, M. S. Alam, Renjish K. Philip, and Sharif Ahmad. "Linseed amide diol/DGEBA epoxy blends for coating applications: preparation, characterization, ageing studies and coating properties." Progress in Organic Coatings 67, no. 2 (2010): 170–179.

37. Boronin, A. I., S. V. Koscheev and G. M. Zhidomirov, "XPS and UPS study of oxygen states on silver," *Journal of Electron Spectroscopy and Related Phenomena* 96, (1998): 43–51.
38. Ferrari, Andrea C. "Raman spectroscopy of graphene and graphite: disorder, electron–phonon coupling, doping and nonadiabatic effects." Solid State Communications 143, no. 1–2 (2007): 47–57.
39. Mallick, P. K. *Processing of Polymer Matrix Composites: Processing and Applications*, CRC Press, Boca Raton, FL, 2017.
40. ASM International. Handbook Committee. *ASM Handbook: Mechanical Testing and Evaluation*. Vol. 8, ASM International, Novelty, OH, 2000.

Section III

Analysis

7 Aging and Corrosion Behavior of Ni- and Cr-Electroplated Coatings on Exhaust Manifold Cast Iron for Automotive Applications

T. Ramkumar
Dr. Mahalingam College of Engineering and Technology

C. A. K. Arumugam
Mepco Schlenk Engineering College

M. Selvakumar
Dr. Mahalingam College of Engineering and Technology

CONTENTS

7.1	Introduction	111
7.2	Materials and Methods	112
7.3	Results and Discussion	113
	7.3.1 Hardness	113
	7.3.2 Aging Behavior	113
	7.3.3 Corrosion Behavior	114
	7.3.3.1 Weight Loss Method	114
	7.3.3.2 Potential Dynamic Polarization	115
7.4	Conclusion	119
References		119

7.1 INTRODUCTION

In the automobile exhaust system, exhaust manifold plays a vital role [1–5]. The exhaust manifold acts as a passage of internally burnt gases from the engine cylinder to the exhaust system. Also the exhaust manifold may get affected by high-temperature gases that are exhausted from the engine. To reduce the high-temperature effects, coating is applied on the exhaust manifold, and in order to avoid this problem, nickel–chromium is coated on the substrate (gray cast iron). The ongoing progress of gas-driven engines for heavy-duty vehicles will further raise the exhaust-gas temperature and make the gas composition more corrosive. The demand was created for both heat and corrosion resistance of the exhaust manifolds. Moreover, materials with high-temperature corrosion resistance and with the ability to withstand the thermal cycling are to be developed. Ferritic ductile cast iron (SiMo51) material is used as a current exhaust manifold that is working at 800°C. At present, many researchers are investigating to enhance the properties of cast iron manifold coated with some dopants such as Nb, Sn, Fe, Ni, and Cr. For exhaust manifolds, one of the most capable approaches to coat the oxidation resistance is electroplating [6].

Recently, surface modification plays a major role in order to enhance the mechanical and corrosion properties. There are numerous techniques employed, but the electroplating technique is extensively used in the manufacturing sectors because of its cost-effectiveness. In a manufacturing sector, this technique is a promising one because of its high-quality coating. Depositing metal-based coatings onto cast iron surface is a budding significance, in order to sustain the cathodic shield of the cast iron substrate. However, Ni and Cr exhibit elevated corrosion resistance. The influence of Cu, Cr, Ni, and Zn coating on cast iron is described by various researchers [6–10]. Therefore, the objective of this study is to coat Ni and Cr on cast iron with different compositions using the electroplating technique. The mechanical and corrosion properties of the coatings are also deliberated in detail.

7.2 MATERIALS AND METHODS

Cast iron with a thickness of 3 mm was purchased from Coimbatore Metal Mart (P) Ltd., Coimbatore, Tamil Nadu, India. Initially, the sample was cut into 100 × 70 × 3 mm using wire electrical discharge machining (WEDM). The substrate was polished using silicon carbide (SiC) emery sheets of various grit sizes such as 800, 1,000, and 1.500 μm. Alumina suspension was used to achieve a mirror-polished surface. The substrate was electroplated with nickel (Ni) and chromium (Cr) as per ASTM B689 and B650. The coating was done with three different compositions: 75% nickel–25% chromium, 80% nickel–20% chromium, and 85% nickel–15% chromium. Primarily, the substrate was degreased with HCl, followed by rinsing with water. Cast iron substrate was taken as cathode, and a metal to be coated (Ni and Cr) was taken as anode. For Ni electroplating, nickel chloride ($NiCl_2 \cdot 6H_2O$), boric acid (H_3BO_3), and nickel sulfamate ($Ni(SO_3NH_2)_2$) were used to prepare the nickel bath. Likewise, chromic acid (CrO_3) and sulfuric acid (H_2SO_4) were used as source materials for chromium bath for the electroplating process. After the electroplating process, the samples were exposed to passivation in nitric acid for 30 min, followed by rinsing with hot water and drying at room temperature [7]. The electroplated samples were cut into 15 × 10 × 5 mm using WEDM. The surface morphology of the coated samples was assessed using SEM. In order to evaluate the hardness of the coated samples, Vickers' hardness test was carried out on the interface and coated layer with an applied load of 0.01 kg with a dwell time period of 10 s.

Aging test was conducted to determine the occurrence of cracks on the surface of the material when it is subjected to a high temperature at different thermal cycles. Gray cast iron (uncoated material) was placed in a furnace of 850°C for 5 min, and it was cooled by air for another 15 min. Thus, one thermal cycle was completed and this process was again repeated for 20 times or cycles. The microstructure of the surface was viewed through scanning electron microscope (SEM). It was found that the cracks are obtained on its surface due to the brittle nature of gray cast iron.

The corrosion behavior of the uncoated and coated samples was studied into two different methodologies: One is the weight reduction method and another one is the polarization technique (Tafel). The sample's dimension of 15 × 10 × 3 mm was immersed in 3.5% HCl solution. The samples were completely immersed in 100 ml of 3.5% HCl solution in a beaker. Before conducting the experiment, the weight of the samples was calculated using electronic weighing balance having accuracy of 0.001 mg. Further, before and after the experimentation, pH and electrical conductance were noted. The whole experiment was conducted at room temperature. Using this process, the corrosion rate was evaluated using the weight loss method.

The corrosion resistance of the coated samples was also monitored through corrosion density (Icorr) and corrosion potential (Ecorr) of the uncoated and coated samples against corrosive environment. Prior to the estimation, the coated cast iron was allowed to be steady in the electrolyte to attain a steady open circuit (OCP). Electrolytic plated samples were taken as the working electrode and the reference electrodes as Ag/AgCl and a platinum wire. According to ASTM G3-14 standard, the corrosion potential (Ecorr) and corrosion current density (Icorr) were evaluated using the Tafel plot [8–10]. The experiments were conducted three times, and the average value was reported. After polarization, the surface morphology of the corroded material was examined using SEM.

7.3 RESULTS AND DISCUSSION

7.3.1 Hardness

Vickers' hardness of the uncoated and coated manifolds is displayed in Figure 7.1. It clearly displays that the hardness of the coated manifolds was increased. The data points of the hardness were collected at various points of the manifolds. The experiments were conducted three times, and then, the average value was reported. The higher hardness was attained at 85 Ni-15 Cr compositions. The enhanced hardness was obtained because Ni-rich phase was coated over the manifolds. Hardness is directly proportional to the refinement of microstructure; it reflects the good bonding between the coating surfaces. It is apparent that the microhardness influences the higher dislocation density in the bare manifolds, because of the variation in their coefficient of thermal expansion.

7.3.2 Aging Behavior

Gray cast iron and coated material were placed in a furnace of 850°C for 5 min. It was suddenly cooled by air for another 5 min; thus, one thermal cycle was completed, and this process was again repeated for five thermal cycles. The microstructure of the

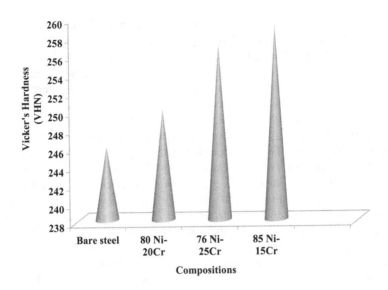

FIGURE 7.1 Vickers' hardness of uncoated and coated samples.

surface was investigated by SEM image. No cracks were identified on the surface of uncoated samples. Hence, it was decided to conduct another 15 thermal cycles for crack identification, but no cracks were found on its surface. Further, the thermal cycle was increased and the crack identification process was carried out for 20 thermal cycles, at which the cracks are clearly visible on the surface. Figure 7.2 displays the microstructure changes of various cycles for bare steel [11].

Figure 7.3 depicts the SEM micrographs of cast iron manifolds of different compositions before and after the thermal cycles. The microstructure of the surface was investigated using SEM. It was found that the cracks are obtained on its surface due to the brittle nature of the gray cast iron [12,13]. A similar test was carried for the coated samples under the same condition, and their microstructure was viewed, whereas no cracks were identified on the surface because nickel and chromium are non-brittle in nature. The presence of nickel and chromium in the coated samples of different compositions (i.e., Ni 75%–Cr 25%, Ni 80%–Cr 20%, and Ni 85%–Cr 15%) reveals the same results as no cracks were observed on the surface.

7.3.3 Corrosion Behavior

The corrosion behavior of the coated manifolds evaluated using the weight reduction method and Tafel exploration is discussed in this section.

7.3.3.1 Weight Loss Method

Table 7.1 shows the weight loss for uncoated manifolds and coated cast iron manifolds (85 Ni-15 Cr). The experiments were conducted for 15 days in an acidic medium (1 mole HCl solution). It was noticed that the weight loss was reduced for coated manifolds. Ni and Cr act as a barrier for cast iron manifolds. But, bare cast iron

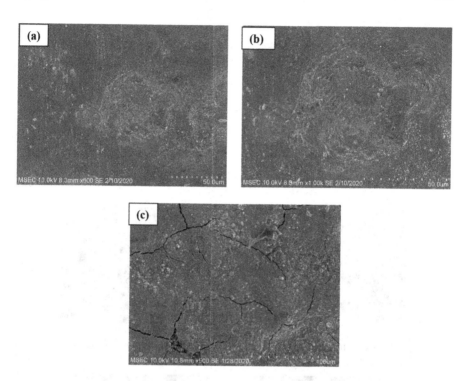

FIGURE 7.2 SEM images of bare steel. a) 5, (b) 15, and (c) 20 cycles.

reacts with HCl strongly and leads to a reduction in weight. However, the weight loss was increased significantly, if the immersion time period of the samples increased. The ability of iron materials has led to excessive corrosion, and hence, it attacked the cast iron very violently.

7.3.3.2 Potential Dynamic Polarization

Figure 7.4 shows the potentiodynamic polarization (Tafel region) for coated and uncoated cast iron samples. From the graph, it is illustrated that for the coated sample of Ni 85%–Cr 15%, the curve shifts toward the positive side compared to the other samples. Using Tafel polarization, the corrosion rate was assessed for all the samples. OCP circuit reveals the thermodynamic parameter, which leads to the tendency of metallic materials to participate in the electrochemical reactions in a chloride medium.

The Tafel plot illustrates the logarithmic relation between current generated between electrochemical cells and electrode potential of a specific material. This plot was generated based on electrochemical reactions between the samples and medium at a controlled atmosphere. However, based on the dip value of the specimen, the plot was generated. More dip value creates less corrosion resistance and vice versa. The uncoated samples have more dip than the coated samples. Ni 85%–Cr 15% exhibits less dip value that leads to high corrosion resistance than the others [13]. As shown in Figure 7.4, the corrosion potential of Ni 85%–Cr 15% coatings in the range of −0.5 to −0.63 V. Uncoated samples have −0.67 V, which is more negative than the coated samples. The maximum

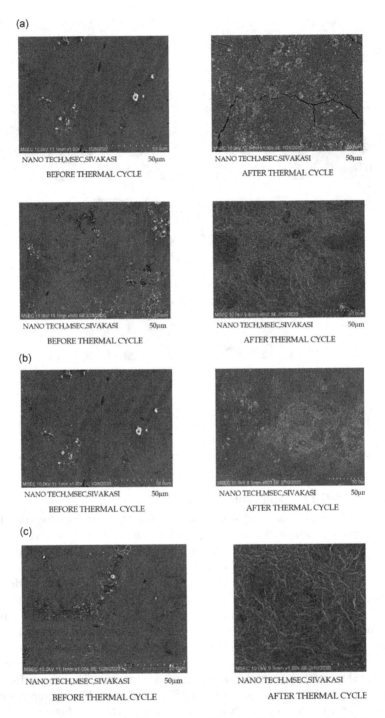

FIGURE 7.3 SEM micrographs of cast iron manifolds of different compositions before and after the thermal cycles. (a) Ni 85%–Cr 15%; (b) Ni 80%–Cr 25%; and (c) Ni 75%–Cr 25%.

TABLE 7.1
Variation of Weight Loss for Bare Cast Iron and Coated Cast Iron Samples

Period (Days)	Bare Cast Iron (g)	Coated Cast Iron (g) Ni 85%–Cr 15%	Difference (g)
Initial weight	19.07	25.39	6.32
1	17.05	25.30	8.25
2	16.90	25.24	8.34
3	16.82	25.20	8.38
4	16.74	25.17	8.43
5	16.62	25.14	8.52
6	16.50	25.11	8.61
7	16.43	25.08	8.65
8	16.34	25.06	8.72
9	16.29	25.00	8.71
10	16.20	24.91	8.71
11	16.10	24.83	8.73
12	16.01	24.78	8.77
13	15.90	24.73	8.83
14	15.72	24.66	8.94
15	15.68	24.58	8.90

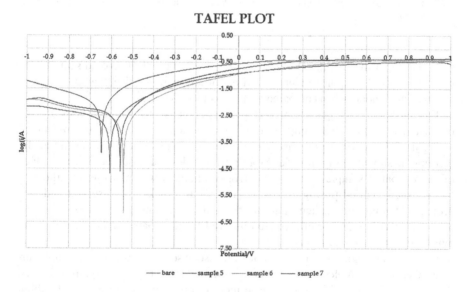

FIGURE 7.4 Tafel plot for uncoated and coated samples.

corrosion potential was observed for the Ni 85%–Cr 15% sample. The corrosion potential is somewhat less for the uncoated sample, the value of which is mentioned before. The corrosion potential deviation was mainly depending on the chemical composition

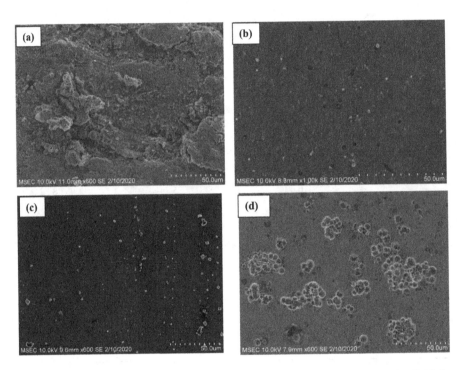

FIGURE 7.5 SEM micrograph of corroded surface. (a) Uncoated cast iron, (b) 85 Ni-15 Cr, (c) 80 Ni-20 Cr, and (d) 75 Ni-25 Cr.

of Ni and Cr of the coating. During the experimentation, for the coated sample with Ni 85% and Cr 15%, polarization is shifting toward more positive potentials; thus, it becomes more noble when compared to the uncoated samples.

For superior investigation of this phenomenon, surface morphology (SEM) of coated and uncoated samples after polarization is essential and presented in Figure 7.5a–d. Uncoated cast iron sample is violently attacked on the acidic medium as shown in Figure 7.5a. More pitting holes are observed on the surface of the uncoated samples. Uniform corrosion damage was observed on the surfaces, usually called pitting corrosion [14]. Small cracks are observed and it is propagated along the surfaces and thus formed the continuous crack, which leads to stress cracking. Minor cracks are observed for the coated samples, which are evidenced in Figure 7.5b–d.

On the other hand, some of the dimples are also observed on the coated sample surfaces. Generally, dimples reduce the corrosion rate. From Figure 7.5d, it is perceived that the Ni 85%–Cr 15% samples have more dimples on the surfaces and exhibit good corrosion resistance than the other samples. From the SEM micrographs, it is clearly visualized that no significant attack is found on the coated surfaces and only minor corrosion is observed [15]. It can be concluded that the uniformity and homogeneity of the Ni 85%–Cr 15% composition are considered to be good candidates for coating on cast iron manifolds.

7.4 CONCLUSION

An intense glossy attractive Ni- and Cr-rich layer is formed on the exhaust manifold cast iron substrate. Vickers' hardness was determined on the cross section of the uncoated and Ni- and Cr-electroplated cast iron samples which shows a significant increase in microhardness of 246 and 259 HV, respectively. Very closely packed Ni and Cr layer deposited on cast iron indicates the enhanced surface property of cast iron. It was also found that the crack was reduced in the Ni- and Cr-coated cast iron manifold when compared to the uncoated manifold. A significant weight reduction was observed on the coated samples compared to the uncoated cast iron manifold. Further, the corrosion resistance was considerably enhanced for the 85 Ni–15 Cr-coated samples. Thus, the nickel–chromium plating has been considered as an excellent corrosion resistance for mass production.

REFERENCES

1. Wanjun He, Rui Hu, Yang Wu, Xiangyu Gao, Jieren Yang (2018), "Mechanical properties of an aged Ni-Cr-Mo alloy and effect of long-range order phase on deformation behavior," *Materials Science and Engineering: A*, 731, 29–35.
2. Yake Wu, Ya Li, Junyong Lu, Sai Tan, Feng Jiang, Jun Sun (2019), "Effects of predeformation on precipitation behaviors and properties in Cu-Ni-Si-Cr alloy," *Materials Science and Engineering: A*, 742, 501–507.
3. J.R. Deepak, V.K. Bupesh Raja, Gobi Saravanan Kaliaraj (2019), "Mechanical and corrosion behavior of Cu, Cr, Ni and Zn electroplating on corten A588 steel for scope for betterment in ambient construction applications," *Results in Physics*, 14, 102437.
4. Yun Xie, Thuan Dinh Nguyen, Jianqiang Zhang, David J. Young (2019), "Corrosion behaviour of Ni-Cr alloys in wet CO_2 atmosphere at 700 and 800°C," *Corrosion Science*, 146, 28–43.
5. Fei Teng, David J. Sprouster, George A. Young, Jia-Hong Ke, Julie D. Tucker (2019), "Effect of stoichiometry on the evolution of thermally annealed long-range ordering in Ni–Cr alloys," *Materialia*, 8, 100453.
6. Biaobiao Yang, Chenying Shi, Jianwei Teng, Xiaojuan Gong, Xianjue Ye, Yunping Li, Qian Lei, Yan Nie (2019), "Corrosion behaviours of low Mo Ni-(Co)-Cr-Mo alloys with various contents of Co in HF acid solution," *Journal of Alloys and Compounds*, 791, 215–224.
7. Wanjun He, Rui Hu, Xiangyu Gao, Jieren Yang (2017), "Evolution of $\Sigma 3^n$ CSL boundaries in Ni-Cr-Mo alloy during aging treatment," *Materials Characterization*, 134, 379–386.
8. Hermann Kirchhöfer, Florian Schubert &Hubertus Nickel (1984), "Precipitation behavior of Ni–Cr–22 Fe–18 Mo (Hastelloy-X) and Ni-Cr-22 Co-12 Mo (Inconel-617) after isothermal aging," *Nuclear Technology*, 66, 139–148.
9. Pingli Mao, Yan Xin, Ke Han (2012), "Anomalous aging behavior of a Ni–Mo–Cr–Re alloy," *Materials Science and Engineering: A*, 556, 734–740.
10. Zhi-yuan Zhu, Yi Sui, An-lun Dai, Yuan-fei Cai, Ling-Li Xu, Ze-xin Wang, Hong-mei Chen, Xing-ming Shao, Wei Liu (2019), "Effect of aging treatment on intergranular corrosion properties of ultra-low iron 625 alloy," *International Journal of Corrosion*, doi: 10.1155/2019/9506401.
11. T. Ramkumar, M. Selvakumar, M. Mohanraj, P. Chandrasekhar, K. Gobi Saravanan (2019), "Effect of TiB addition on corrosion behaviour of Titanium composites under Neutral Chloride solution," *Transactions of the Indian Ceramic Society*, 78(3), 155–160.

12. Yubi Zhang, Xiaoyang Hu, Changrong Li, Weiwei Xu, Yongtao Zhao (2017), "Composition design, phase transitions of a new polycrystalline Ni-Cr-Co-W base superalloy and its isothermal oxidation dynamics behaviors at 1300°C," *Materials & Design*, 129, 26–33.
13. Heng Zhang, Yuan Liu, Xiang Chen, Huawei Zhang, Yanxiang Li (2017), "Microstructural homogenization and high-temperature cyclic oxidation behavior of a Ni-based superalloy with high-Cr content," *Journal of Alloys and Compounds*, 727, 410–418.
14. B. Gao, L. Wang, Y. Liu, X. Song, X. Song, A. Chiba (2019), "High temperature oxidation behaviour of γ'-strengthened Co-based superalloys with different Ni addition," *Corrosion Science*, 157, 109–115.
15. T. Ramkumar, M. Selvakumar, R. Vasanthshankar, A.S. Sathishkumar, P. Narayanasamy, G. Girija (2018), "Rietveld refinement of powder X-ray diffraction, microstructural and mechanical studies of magnesium matrix composites processed by high energy ball milling," *Journal of Magnesium and Alloys*, 6(4), 390–398.

8 Experimental Evaluation of Wear and Coefficient of Frictional Performance of Zirconium Oxide Nanoparticle–Reinforced Polymer Composites for Gear Applications

S. Sathees Kumar and B. Sridhar Babu
CMR Institute of Technology

CONTENTS

8.1 Introduction 122
    8.1.1 Polyamides 122
    8.1.2 Polyamide 6 123
8.2 Chemical Composition of PA6 123
    8.2.1 Features of PA6 124
    8.2.2 Applications of PA6 124
    8.2.3 Fillers 125
    8.2.4 Materials Used 125
    8.2.5 Preparation of PA6 Test Specimen 125
8.3 Various Stages of Preparation of PA6 Composite Test Specimen 126
    8.3.1 Stage 1: Preheating 126
    8.3.2 Stage 2: Cooling 126
    8.3.3 Stage 3: Stirring/Blending 126
    8.3.4 Stage 4: Molding 126
8.4 Proportions of Fabricated PA6 Composite Test Specimen 126
8.5 Tribological Properties of PA6 Composites 126
    8.5.1 Friction and Wear Rate 126
    8.5.2 Analysis of Tribological Execution of PA6 and PA6 Composites 129
8.6 Coefficient of Friction for PA6 and PA6 Composites 129
8.7 Rate of Wear 131
8.8 Scanning Electron Microscopy Study of ZrO_2/PA6 Composites 133

8.9 Fabrication of Gear 135
8.10 Conclusions.. 136
References... 136

8.1 INTRODUCTION

Polymer composites are as a rule progressively utilized in the plastic industry due to their great qualities and low densities (Nie et al., 2010). Particularly short fiber–strengthened thermoplastic polymer composites are broadly utilized in many fields, for example, airplane, aviation, and car industry (Molnàr et al., 1999; Botelho et al., 2003). Among the thermoplastic polymers, polyamide 6 (PA6) has become a solid contender grid inferable from its great warm security, low dielectric steady, and high elasticity (Botelho and Rezende, 2006; Li, 2008).

The primary drawbacks of polyamides (PAs) is their high moisture retention, which brings down the checked on mechanical properties and dimensional solidness (Ebewele, 2000; Strong, 2006). Additionally, research shows that PA materials lose their mechanical properties at high temperatures (Mao, 2007). To overcome the above limitation, some researchers have reported that the mechanical and wear resistances can be improved when the polymers are reinforced with fillers (Sung and Suh, 1979; Li et al., 2000; Chen et al., 2008; Kowandy et al., 2008). Usually, adding fillers to the polymers will increase some properties like mechanical, tribological, and thermal stabilities of the polymer (Unal and Mimaroglu, 2012). Fillers are fundamentally basic inorganic mineral powders added to improve handling, rigidity, and dimensional constancy as expressed (Brydson, 1966).

The various reinforcements used in PA materials are graphite, carbon, wax, polytetraflouroethylene, polyethylene terephthalate, silica, carbon nanotube, carbon fiber, and high-density polyethylene. In this experiment, to improve the tribological properties of the pure PA6, it is decided to reinforce with the filler zirconium dioxide (ZrO_2) in different weight proportions. ZrO_2 is one of the most capable nanoparticles utilized in anticorrosion coatings. Nanometric ZrO_2 particles are an innovatively significant class of materials with a wide scope of uses. ZrO_2 exposes amazing properties, for example, better strength, maximum break toughness, good wear resistance, high hardness, and great substance opposition. Not many works were carried out in the past about the impact of ZrO_2 on the properties of thermoplastics. A few studies have detailed that ZrO_2 nanoparticles indicated better biocompatibility when compared with different nanomaterials, including ferric oxide, titanium dioxide, and zinc oxide

In concurrence with these outcomes, others have announced that ZrO_2 nanoparticles could prompt gentle (Karunakaran et al., 2013) or no cytotoxic impacts.

8.1.1 POLYAMIDES

A point-by-point investigation of the impact of these boundaries on the composite properties reveals that PAs are a significant gathering of the thermoplastic polycondensates. The amide gathering occurs by the polymerization of lactams (polylactams) or by the buildup of diamines with nylons. Hardly any creators expressed

Evaluation of Wear and Coefficient

that PAs continually pull in more extensive intrigue on account of their interesting mechanical, thermal, and morphological properties. PAs are notable for their fantastic mechanical performances. The two substantial sorts of PAs are polyamide 66 (PA66) and PA6. PA66 is made out of two basic units, each with six carbon atoms, in particular the residues of hexamethylenediamine and adipic acid (Hatke et al., 1991; Spiliopoulos and Mikroyannidis 1998; Liaw et al., 2003).

PA6 or poly 6-caprolactam, another significant polymer, involves a single structural unit, to be specific the clear residue of 6-aminocaproic acid. PAs do uncover an inclination to creep under applied load. Likewise, the properties of PAs are extensively influenced by moisture. PAs have a few advantages over different classes of polymers. The thermoplastic polymers are a class of engineering materials that develop commercial outcome because of their simplicity of assembling and great thermomechanical properties (Benaarbia et al., 2014).

8.1.2 POLYAMIDE 6

PA6 is a polymer created by Paul Schlack at Interessen—Gemeinschoft Farben to reproduce the properties of PA66 without disregarding the patent of its creation. In contrast to most different PAs, PA6 is not a buildup polymer, rather it is encircled by ring-opening polymers. PA6 begins as a crude caprolactam. As caprolactam has 6 carbon molecules, it has the name Nylon 6. When caprolactam is warmed at 533K in an inert environment of nitrogen for around 4–5 h, the ring breaks and encounters polymerization. By at that point, the liquid mass has encountered the spinners to shape fiber of Nylon 6.

8.2 CHEMICAL COMPOSITION OF PA6

PAs are a gathering of thermoplastic polymers containing amide bunches (–CONH) in the fundamental chain. They are famously known as Nylon 6. PA 6 [NH–$(CH_2)_5$–CO] is produced using ε-caprolactam. It is framed by ring-opening polymerization of ε-caprolactam.

The chemical composition of PA6 is shown in Figure 8.1.

Alternative name: poly-ε-caproamide (Source: shodhganga.inflibnet.ac.in)
Trade names: Capron, Ultramid, Nylatron (Source: shodhganga.inflibnet.ac.in)
Class: Aliphatic polyamides (Source: shodhganga.inflibnet.ac.in)

$$H_3N^+ (CH_2)_5 CO \xrightarrow[-H_2O]{\Delta} -NH(CH_2)_5 C(=O) - \left[NH(CH_2)_5 C(=O) \right] - NH(CH_2)_5 C(=O)$$

6 - aminohexanoic acid   PA6 a Polyamide

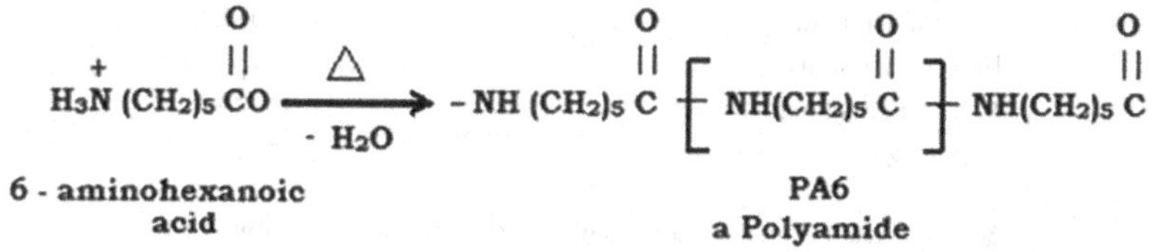

FIGURE 8.1 Chemical composition of PA6. (Paula Yurkanis Bruice, 2006.)

8.2.1 Features of PA6

PA6 strands are intense and have high rigidity, just as versatility and radiance. They are wrinkle confirmation. They are additionally resistant to abrasion and synthetic elements, for instance, acids and solvable bases (Pogacnik and Kalin, 2012). Actual properties of PA6 are satisfactory wear and rough security, low coefficient of grating, high adaptability, high caliber and solidness got together with extraordinary impact opposition. PA6 is a crystalline thermoplastic that has a low thermal coefficient with linear expansion, has a low coefficient with thermal development, and is exceptionally sensitive to dampness.

PA6 has become a sturdy competitor network, attributable to its great thermal constancy, low dielectric consistent, and superior elasticity (Karsli and Aytac, 2013). The general features of PA6 are given in Table 8.1.

8.2.2 Applications of PA6

PAs are prospective thermoplastic materials used for innumerable purposes, which is inferred from their amazing expansive properties (Li et al., 2013). PA6 is utilized as a string in bristles for toothbrushes, in surgical stitches, and in acoustic and traditional instruments, including guitars, violins, violas, and cellos. It is likewise utilized in the production of an enormous assortment of strings, ropes, fibers, nets, and tire ropes, as well as in hosiery and weaved articles of clothing. In an automobile industry, it is employed in wire and link jacketing, cooling fans, air admission, turbo air channels, valve and engine covers, brake and force guiding repositories, gears for windshield wipers, speedometers, and numerous other automotive parts (Mallick, 2007). It is utilized as gear and bearing materials as a result of their equalization in quality, hardness, and sturdiness and in view of their great friction characteristics (Bermudez et al., 2001).

TABLE 8.1
General Features of PA6

Property	Value
Glass transition temperature	40°C
Melting temperature	220°C
Density (crystalline) at 25°C	1.23 g/cc
Density (amorphous) at 25°C	1.084 g/cc
Tensile strength	35–230 MPa
Tensile modulus	2,000–3,000 MPa
Flexural strength	40–230 MPa
Flexural modulus	1,800–2,414 MPa
Heat deflection temperature	58°C
Surface hardness (shore D)	79

Source: Data from designerdata.nl/materials/plastics/thermo-plastics/polyamide-6.

8.2.3 Fillers

Generally, fillers are considered as added substances, in view of their unfavorable geometrical characteristic and surface area or surface chemical composition. They can just moderately update the modulus of the polymer, while the quality remains unaltered or even decreased. Particulate filler-fortified composites appear to offer various recompenses over neat resin matrices, incorporating heightened stiffness, strength- and dimension-dependent qualities, upgraded durability or impact, improved heat distortion temperature, expanded mechanical damping, diminished penetrability to gases and fluids, adjusted electrical properties, and diminished prices (Nielsen and Landel, 1994; Kumar and Wang, 1997). The particulate filler for polymer composite systems is open in a couple of sizes and shapes, including sphere, cubic, platelet, or some other predictable or uneven geometry (Katz, 1998). Aside from the mechanical properties, various qualities of the material can be improved because of development of the filler (Konieczny et al., 2013).

8.2.4 Materials Used

PA6: Nylon 6 (coded as PA6): pellet size—3 mm, density—1.40 g/cm^3, physical state—white, and appearance—chips. Similarly, zirconium dioxide: particle size—45–55 nm range and density—5.22 g/cm^3. PA6 and ZrO$_2$ fillers are shown in Figure 8.2.

8.2.5 Preparation of PA6 Test Specimen

The melting temperature of PA6 is 220°C. As a result, PA6 pills are filled in the injection molding equipment and they are directly molded by the injection molding device. At this heat (210°C), PA6 is melted and converted into a molten state. The obtained liquid PA6 is passed from the injection molding device to a preheated die. The preheated die is used to fabricate the specimens as per the following dimensions: 35 mm length and 25 mm diameter tribological experimental tests. The obtained specimens are machined as per ASTM test standards for tribological tests.

FIGURE 8.2 PA6 and fillers (a) PA6 (b) ZrO$_2$.

8.3 VARIOUS STAGES OF PREPARATION OF PA6 COMPOSITE TEST SPECIMEN

8.3.1 Stage 1: Preheating

Fillers (ZrO_2) have high melting temperatures and subsequently, they will not blend with PA6 promptly. Thus, fillers should be preheated under the liquefying temperatures. Here, the ZrO_2 is warmed under the liquefying heat right throughout in a muffle furnace up to 2,670°C.

8.3.2 Stage 2: Cooling

As a rule, the liquefying hotness of PA6 is 210°C. If it rapidly mixes with the filler, it will fire. Hence, to subsequently attain the heat of 2,670°C, the filler was ventilated at air temperature for 40–50 min. By this cooling procedure, the filler transforms to form ash.

8.3.3 Stage 3: Stirring/Blending

The cooled filler was mixed with PA6 pills by the blending process. The blending procedure is done in a 75-mm-diameter and 100-mm-height cast iron crucible.

8.3.4 Stage 4: Molding

The mixed material is liquefied at 210°C in the injection molding device. At this heat, PA6 liquefies and reaches the molten state.

The melted PA6 is delivered from the injection molding device to 75°C heated die. Now, the prewarming of the die is useful for smooth streaming of materials. The fabrication of specimen process and fabricated specimens are exhibited in Figures 8.3 and 8.4, respectively.

8.4 PROPORTIONS OF FABRICATED PA6 COMPOSITE TEST SPECIMEN

PA6 90 wt.% with filler 10 wt.%, PA6 80 wt.% with filler 20 wt.%, PA6 70 wt.% with filler 30 wt.%, and PA6 60 wt.% with filler 40 wt.%. Temperature varies in each phase of sample preparation, and particulars of manufactured test samples of PA6 and PA6 composites appear in Tables 8.2 and 8.3 individually.

8.5 TRIBOLOGICAL PROPERTIES OF PA6 COMPOSITES

8.5.1 Friction and Wear Rate

The dry sliding wear qualities of the mixture were explored using a pin-on-plate analyzer as indicated by ASTM G99 standard. The size of the test is 12 mm width and 30 mm length. The outsides of both the example and the plate were scrubbed

Evaluation of Wear and Coefficient 127

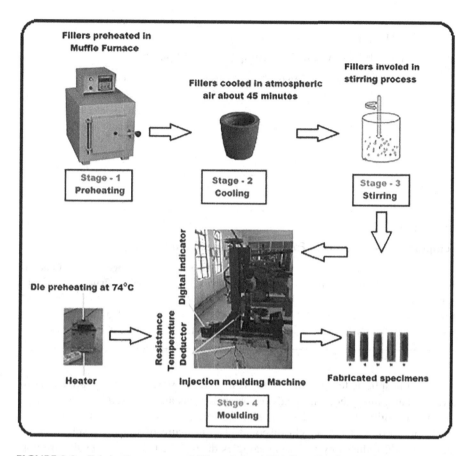

FIGURE 8.3 Fabrication process of filler-reinforced PA6 test specimens.

FIGURE 8.4 Fabricated PA6 and ZrO$_2$/PA6 composite specimens.

with sensitive acetone-assimilated paper before the test. The testing rate is had in the range of 1,000, 1,500, and 2,000 rpm. At that point, the weight is attached to the scope of 10N, 20N, 30N, and 40N. The tribological qualities of five samples are tried by pin-on-disc tester over various dry abrasion settings. Tribometer designs since the contact occurring between two-level surfaces forestall an enormous contrast in the district of contact between the polymer and the harder counter ace on account of wear and creep as explored (Palabiyik and Bahadur, 2002).

TABLE 8.2
Details of PA6 and PA6 Composite Test Specimens

Specimen Code	PA6 (wt.%)	ZrO_2 (wt.%)	Remarks
S1	100	-	Raw PA6
S2	90	10	ZrO_2-reinforced
S3	80	20	PA6 specimen
S4	70	30	
S5	60	40	

TABLE 8.3
Temperature Variations in Each Stage of Specimen Preparation

Test Specimen	Stage 1 Preheating Temp. of Fillers (°C)	Stage 2 Cooling Temp. (°C)	Stage 3 Stirring Temp. of PA6 (°C)	Stage 4 Molding Temp. of PA6 (°C)	Code No of Test Specimen
PA6	-	32	-	210	S1
ZrO_2/PA6	2670	32	122–128	210	S2–S5

The atmospheric temperature in the test center is 31°C and 46% relative humidity. The basic and preceding loads of the specimens are assessed by exploiting an electronic digital balance. The division between the basic and remaining loads is the proportion of the weight discount. The measure of wear is resolved as far as the weight reduces by weighing samples during the test. Wear results are normally gained by operating a test on chosen sliding distance and chosen estimations of load and speed. A few papers are tended to as of now the wear execution of

TABLE 8.4
Tribological Testing Specifications of ZrO_2-Strengthened PA6 Composites

Constraints	Testing Conditions
Applied load	10N, 20N, 30N, and 40N
Material of disc	EN31 steel
Speed of disc	1,000–2,000 rpm
Materials of pin	ZrO_2-reinforced PA6 (S2–S5)
Surface condition	Dry
Diameter of track	100 mm
Distance of sliding	1,000 m

TABLE 8.5
Tribological Functioning of PA6 and PA6 Mixtures at Different Loads

Specimen Code	Friction Coefficient with Loads				Wear Rate $\times 10^{-3}$ (mm^3/Nm) With Loads			
	10N	20N	30N	40N	10N	20N	30N	40N
S1	0.32	0.35	0.37	0.42	7.2	7.5	7.8	8.2
S2	0.31	0.33	0.35	0.37	6.3	6.5	6.8	7.0
S3	0.32	0.34	0.36	0.38	6.5	6.7	6.9	7.1
S4	0.34	0.36	0.38	0.39	6.7	6.9	7.1	7.4
S5	0.28	0.30	0.32	0.34	6.1	6.4	6.7	6.9

modified thermoplastics and thermosets (Sung and Suh, 1979). Here, the accompanying requirements and the trail specifications that appear in Table 8.4 are used as a portion of the tribometer.

8.5.2 ANALYSIS OF TRIBOLOGICAL EXECUTION OF PA6 AND PA6 COMPOSITES

From the trial consequences of tribological tests, coefficient of contact and wear rate results are attained. Tribological functioning of PA6 and PA6 composites is expressed in Table 8.5.

8.6 COEFFICIENT OF FRICTION FOR PA6 AND PA6 COMPOSITES

Figure 8.5a–d shows the variety of friction for PA6 (S1) and ZrO_2/PA6 composites (S2–S4) for various loads. The drop in friction with an expansion in weight is ascribed as a result of greatest sliding speeds at various weight percentages of reinforcement. The ZrO_2-strengthened PA6 composite materials have an inferior coefficient of friction when compared with raw PA6. Be that as it may, the analysis illustrates that the composite material shows a trivial variation in the friction coefficient with the variation in load and sliding speed. In Figure 8.5, ZrO_2-strengthened PA6 composite materials have a lesser friction coefficient when compared with raw PA6. From these outcomes, it is distinguished that the PA6 with 30 wt.% (S5) has the least friction at all loads. It is presumed that the expansion of applied load demonstrates a substantial increment in the friction coefficient as shown in Figure 8.5a–d. Taking every one of these results, it is understood that the event of ZrO2 performs to make ramifications for the frictional coefficient of PA6. Clearly, the friction coefficient builds at first to a higher value because of the new rough material and as the procedure proceeds, it almost stays same for the whole test. It is likewise noticed that the coefficient of friction diminishes, when the filler substance increments. The assessment of the frictional coefficient is comfortable loads in crude PA6 and invigorated PA6 composites.

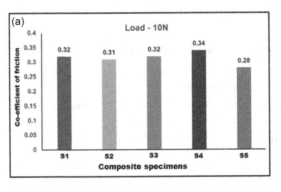
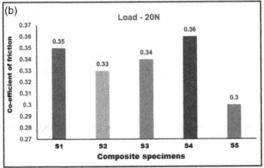
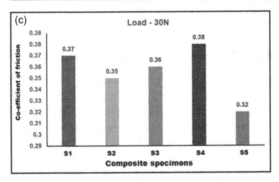
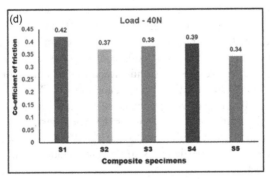

FIGURE 8.5 Coefficient of friction for various load conditions.

8.7 RATE OF WEAR

The wear rate is predicted making use of the given formula:

$$\text{Wear rate} = (\Delta m/\rho L d) \times 10^3 \, \text{mm}^3/\text{Nm} \qquad (8.1)$$

Where Δm is the loss of mass in grams, ρ is the test material density in g/cm³, L is the applied load in Newton, and d is the distance of sliding in meters. The readings of the diagram are plotted at normal time interval of 5 min.

The sample surface is estimated utilizing an electronic balance with the precision of ±0.01 mg. Figure 8.5. Coefficient of friction for various loads of specimens: (a) 10N (b) 20N (c) 30N, and (d) 40N. During the primary tests, the exteriors of both the samples and the steel matching part are coarse and accordingly solid intercorrelations involving the outside bring about high friction coefficient. As the wear procedure proceeds, the coarse contours of the steel matching part and the samples are softened. With an expansion in sliding speed, the frictional hotness is significantly high, and the extra wear fragments follow on the sample. Figure 8.6 shows the wear attributes of PA6 and ZrO_2/PA6 composites under 10N, 20N, 30N, and 40N loads. Applied weight is a few of the highly considerable variables influencing the wear of the composites. The wear rate of the composite material has critical variations when compared with raw PA6 material at entire loads. In Figure 8.6, ZrO_2-filled PA6 composites (S2–S5) show remarkable weakening in wear rate. In Figure 8.6a, the particular wear rate is moderately high at the load of 10N. It happens due to the less quantity of fillers. In tribological test, though the sample interacts with the disc, wear is created on the interaction exterior of the sample. The abrasion wear resistance is significantly expanded at higher loads in light of the fact that the greater part of the filler particles infiltrate into the PA6 polymer surface and furthermore ensure the wear debris in the wear test. The aftereffect of Figure 8.6 shows that the PA6-reinforced 40 wt.% ZrO_2 (S5) has diminished wear rate clearly when compared with raw PA6 and different composites. ZrO_2 particles firmly bond with a primary material. They ensure the outside opposed to extreme destructive activity in the counterface. The wear of the composites lessens with increment in level of ZrO_2 particles as signified in Figure 8.5. However, the frictional coefficient of composites weakens with an expansion in the substance of ZrO_2 particles as determined in Figure 8.6. Figure 8.6 exposes that the wear rate is in the limit of 7.0×10^{-3} and 6.3×10^{-3} mm³/Nm for PA6 with 5 wt.% ZrO_2 composite, 7.1×10^{-3} and 6.5×10^{-3} mm³/Nm for PA6 with 10 wt.% ZrO_2 mixture, and the wear rates are in the scope of 7.4×10^{-3} and 6.7×10^{-3} mm³/Nm for PA6 with 30 wt.% ZrO_2 composite.

As a last fact, the normal wear rate is in the limit of 6.9×10^{-3} and 6.1×10^{-3} mm³/Nm for PA6 with 40 wt.% ZrO_2 composite. Be that as it may, the normal wear rate is 8.2×10^{-3} and 7.2×10^{-3} mm³/Nm for raw PA6. Subsequently PA6 with 40 wt.% ZrO_2 mixture produced high wear obstruction and minimal frictional coefficient on an exterior level when compare with PA6 and different composites. Generally, the ZrO_2 particles have inside and out upgraded hardness and wear opposition. In the same manner, PA6-reinforced 40 wt.% ZrO_2 had a less coefficient of friction and discrete wear rate 6.9×10^{-3} mm³/Nm. Regardless, the result of Figure 8.6 shows that

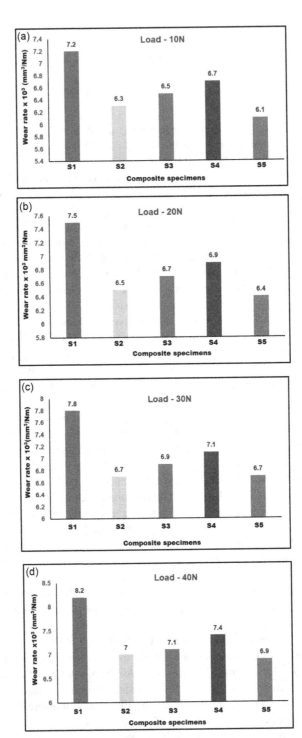

FIGURE 8.6 Wear rate for various load conditions.

the PA6-reinforced 40 wt.% ZrO_2 wear rate was lessened in clearly analyzed PA6 and various mixtures. In the tribological test, the sample connected with the disc/plate, the fundamental credits of warmth and wear has made on the communication and outside of the sample. ZrO_2 elements have high stiffness and superior thermal conductivity. In this way, when fortified with PA6, the attributes of PA6 have been improved. Here, the components disseminated on the PA6 test collaborate with the disc/plate the highlights of hotness and wear have protected from the communication surface of the disc/plate. Thus, it has created extreme wear opposition on the surface of the sample.

8.8 SCANNING ELECTRON MICROSCOPY STUDY OF ZrO_2/PA6 COMPOSITES

Figure 8.7a and b reveals the morphology depictions of raw PA6 test sample of when-worn sides. Few destroyed elements and plastic distortion were discovered in the before sported surfaces of raw PA6 in Figure 8.7a. In Figure 8.7b, the wedge course of action and plastic bending were made on the faces of additional voids and little extension parts were discovered in the after-worn surfaces of PA6. Figure 8.7c and Figure 8.7d ZrO_2 particles were dependably scattered in PA6. At that point, PA6 with 10 wt.% ZrO_2 deciphers the plastic contortion and voids were made less contrasted and PA6 from 20 wt.% ZrO_2. Broken particles, little scope breaks, and wedge headway were not uncovered outwardly of PA6 with 10 wt.% ZrO_2. Figure 8.7e shows the voids strategy and damaged particles of PA6 with 20 wt.% ZrO_2. Figure 8.7f shows the space development, plastic mutilation, and damaged elements of PA6 with 30 wt.% ZrO_2, the clarification of that ZrO_2 particles were not dependably scattered in the PA6. Figure 8.7g and Figure 8.7h show the depiction of 30 wt.% ZrO_2, this portrait obviously delineates the identical spread of ZrO_2 elements in PA6. In the interim, plastic bending and damaged particles were surrounded by the before-worn surfaces. After-worn surfaces of 20 wt.% ZrO_2 reveal miniaturized scale breaks game plan on the surfaces of PA6 (Figure 8.7h). Figure 8.7i shows the space enlargement, plastic bending, and damaged elements of PA6 with 30 wt.% ZrO_2. Figure 8.7j the expansion considering the way that after wear of the PA6 surface, they showed the holding of ZrO_2 components is not correspondingly spreading at a better place on top. These pictures disclose the extension of ZrO_2 with PA6, and the permeability the ZrO_2 material has in the PA6. Wear bearing is also apparent in Figure 8.7j (Sathees Kumar and Kanagaraj 2016a, b, c, d, 2017). It shows that better wear happened on the PA6 with 40 wt.% ZrO_2. The all-around utilized appearances of the composite were normally smooth and there had all the stores of being continuously joined wear wreckages arranged along the sliding track.

It is seen that ZrO_2 strengthened with PA6 surfaces as a general rule wear even more quickly. The more degree of vulnerable of ZrO_2 is blended in with PA6 as besides prevalent as the past plan. It is perceived that the dispersing state of ZrO_2 in the PA6 substances is sensibly consistent in the 30 wt.% mixtures. In summation, the even flow of the ZrO_2 elements in the microarrangement of the PA6 is the foremost careful influence for the upgrading in the tribological attributes.

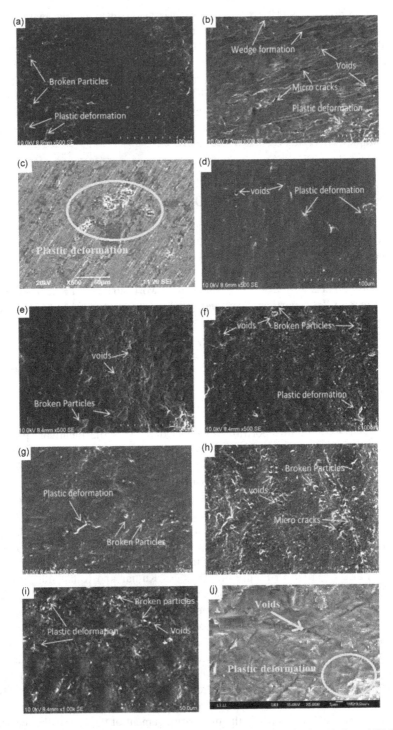

FIGURE 8.7 Scanning electron microscopy images of PA6 and ZrO$_2$-reinforced PA6 composites (a, c, e, g, and i before wear and b, d, f, h, and j after wear).

TABLE 8.6
Specifications of Fabricated Gear

Nomenclature	Values
Module	3 mm
No. of teeth	36
Pitch diameter	108 mm
Tooth depth	5.5 mm
Bottom clearance	0.5 mm
Addendun	2.5 mm
Dedendum	3.0 mm
Working depth	5.0 mm
Tooth whole depth	5.5 mm
Fillet radius	0.8 mm

8.9 FABRICATION OF GEAR

After completion of tribological behavior experiment, the results showed that the 40 wt.% of ZrO_2 attained better tribological attributes. To substantiate the results, the 40 wt.% ZrO_2 + 60 wt.% PA6 composite was fabricated by the following specifications of gear hobbing machine as exposed in Table 8.6.

This type of composite gear can be used instead of PA6 gears for textile industries, automobiles, and numerous engineering applications (Figure 8.8).

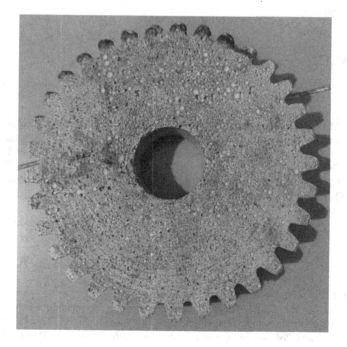

FIGURE 8.8 Fabricated ZrO_2/PA6 composite gear.

8.10 CONCLUSIONS

In this investigation, PA6 mixed with ZrO_2 composites were set up by injection molding device. The impact of ZrO_2 substance of the tribological behavior of PA6 material was assessed from atmospheric temperature. From the investigational outcomes, the subsequent assumptions were made:

- The impacts of ZrO_2 particles on the wear conduct of PA6 polymers were examined. The 40 wt.% ZrO_2 particles are ready to progress the wear obstruction of the PA6 polymer composites. The ZrO_2 particles appeared to be more efficient in raising the wear opposition when compared with that of PA6 polymer.
- From the experimental work, the friction coefficient downsized up to 25% for PA6+40 wt.% ZrO_2 composites when compared with PA6.
- Addition of 40 wt.% measure of ZrO_2 diminished the frictional coefficient of behavior and enhanced the wear opposition of the PA6 matrix. The wear opposition significantly upgraded up to 18.82% for PA6+30 wt.% ZrO_2 composites related to PA6.
- Scanning electron microscopy analysis proved that the even dispersion of ZrO_2 fillers in PA6 matrix enhanced the tribological attributes of polymer composites.
- Substantiate the ZrO_2/PA6 results, and the PA6 composite gear was fabricated.
- These types of composites can be useful for high wear rate and high thermal conductivity in automobile, textile, and numerous engineering applications.

REFERENCES

Benaarbia, A., Chrysochoos, A., & Robert, G. 2014, Kinetics of stored and dissipated energies associated with cyclic loadings of dry polyamide 6.6 specimens. *Polymer Testing*, 34, 155–167.

Bermudez, M.D., Carrion-Vilches, F.J., Martínez-Mateo, I., & Martínez-Nicolas, G. 2001, Comparative study of the tribological properties of polyamide 6 filled with molybdenum disulfide and liquid crystalline additives. *Journal of Applied Polymer Science*, 81, 10, 2426–2432.

Botelho, E.C., Figiel, L., Rezende, M.C., & Lauke, B. 2003, Mechanical behavior of carbon fiber reinforced polyamide composites. *Composites Science and Technology*, 63, 1843–1855.

Botelho, E.C., & Rezende, M.C. 2006, Monitoring of carbon fiber/polyamide composites processing by rheological and thermal analyses. *Polymer-Plastics Technology and Engineering*, 45, 61–69.

Bruice, P. Y. (2006). Química orgânica: Paula Yurkanis Bruice. Traduzido por Debora Omena Futuro... [et al.]. Pearson Prentice Hall.

Brydson, J.A. 1966, *Plastic Materials*, Van-Nostrand, New York.

Chen, Z., Mo, X., He, C., & Wang, H. 2008, Intermolecular interactions in electrospun collagen–chitosan complex nanofibers. *Carbohydrate Polymers*, 72, 3, 410–418.

Ebewele, R.O. 2000, *Polymer Science and Technology*, CRC Press, Boca Ratón, FL.

Hatke, W., Schmidt, H.W., & Heitz, W. 1991, Substituted rod-like aromatic polyamides: Synthesis and structure-property relations. *Journal of Polymer Science Part A: Polymer Chemistry*, 29, 10, 1387–1398.

Karsli, N.G., & Aytac, A. 2013, Tensile and thermomechanical properties of short carbon fiber reinforced polyamide 6 composites. *Composites Part B: Engineering*, 51, 270–275.

Katz, H.S. 1998, *Particulate Fillers, Handbook of Composites*, Chapman & Hall, New York.

Konieczny, J., Chmielnicki, B., & Tomiczek, A. 2013, Evaluation of selected properties of PA6-copper/graphite composite. *Journal of Achievements in Materials and Manufacturing Engineering*, 60, 1, 23–30.

Kowandy, C., Richard, C., & Chen, Y.M. 2008, Characterization of wear particles for comprehension of wear mechanisms: Case of PTFE against cast iron, *Wear*, 265, 11, 1714–1719.

Kumar, S., & Wang, Y. 1997, *Fibers Fabrics and Fillers, Composites Engineering Handbook*, Marcel Dekker Inc., New York.

Li, J. 2008, Interfacial studies on the O3 modified carbon fiber-reinforced polyamide 6 composites. *Applied Surface Science*, 255, 2822–2824.

Li, Z., Xiao, P., Xiong, X., & Huang, B.Y. 2013, Preparation and tribological properties of C fiber reinforced C/SiC dual matrix composites fabrication by liquid silicon infiltration. *Solid State Sciences*, 16, 6–12.

Liaw, D.J., Chen, W.H., & Huang, C.C. 2003, *Polyimides and Other High Temperature Polymers,* Mittal, Kuala Lumpur.

Mallick, P.K. 2007, *Fiber-Reinforced Composites: Materials, Manufacturing, and Design,* CRC Press, Boca Raton, FL.

Mao, K. 2007, A new approach for polymer composite gear design. *Wear*, 262, 3, 432–441.

Molnár, S., Rosenberger, S., Gulyás, J., & Pukánszky, B. (1999). Structure and impact resistance of short carbon fiber reinforced polyamide 6 composites. Journal of Macromolecular Science—Physics, 38(5–6), 721–735.

Nie, W.Z., Li, J., & Zhang, Y.F. 2010, Tensile properties of surface treated carbon fibre reinforced ABS/PA6 composites. *Plastics Rubber and Composites*, 39, 1, 16–20.

Nielsen, L.E., & Landel, R.F. 1994, *Mechanical Properties of Polymers and Composites,* Marcel Dekker Inc., New York.

Palabiyik, M., & Bahadur, S. 2002, Tribological studies of polyamide 6 and high-density polyethylene blends filled with PTFE and copper oxide and reinforced with short glass fibers. *Wear*, 253, 3, 369–376.

Pogacnik, A., & Kalin, M. 2012, Parameters influencing the running-in and long-term tribological behaviour of polyamide (PA) against polyacetal (POM) and steel. *Wear*, 290, 140–148.

Sathees Kumar, S., & Kanagaraj, G. 2016a, Evaluation of mechanical properties and characterization of silicon carbide–reinforced polyamide 6 polymer composites and their engineering applications. *International Journal of Polymer Analysis and Characterization*, 21, 5, 378–386.

Sathees Kumar, S., & Kanagaraj, G. 2016b, Experimental investigation on tribological behaviours of PA6, PA6-reinforced Al_2O_3 and PA6-reinforced graphite polymer composites. *Bulletin of Materials Science*, 39, 6, 1467–1481.

Sathees Kumar, S., & Kanagaraj, G. 2016c, Investigation of characterization and mechanical performances of Al_2O_3 and SiC reinforced PA6 hybrid composites. *Journal of Inorganic and Organometallic Polymers and Materials*, 26, 4, 788–798.

Sathees Kumar, S., & Kanagaraj, G. 2016d, Investigation on mechanical and tribological behaviors of PA6 and graphite-reinforced PA6 polymer composites. *Arabian Journal for Science and Engineering*, 41, 11, 4347–4357.

Sathees Kumar, S., & Kanagaraj, G. 2017, Effect of graphite and silicon carbide fillers on mechanical properties of PA6 polymer composites. *Journal of Polymer Engineering*, 37, 6, 547–557.

Spiliopoulos, I.K., & Mikroyannidis, J.A. 1998, Soluble phenyl-or alkoxyphenyl-substituted rigid-rod polyamides and polyimides containing m-terphenyls in the main chain. *Macromolecules*, 31, 4, 1236–1245.

Strong, B.A. 2006, *Plastics: Materials and Processing*, Prentice Hall, Upper Saddle River, NJ.

Sung, N.H., & Suh, N.P. 1979, Effect of fiber orientation on friction and wear of fiber reinforced polymeric composites. *Wear*, 53, 1, 129–141.

Unal, H., & Mimaroglu, A. 2012, Friction and wear performance of polyamide 6 and graphite and wax polyamide 6 composites under dry sliding conditions. *Wear*, 289, 132–137.

9 Comparative Numerical Analyses of Different Carbon Nanotubes Added with Carbon Fiber–Reinforced Polymer Composite

R. Vijayanandh and G. Raj Kumar
Kumaraguru College of Technology

P. Jagadeeshwaran
Rajalakshmi Institute of Technology

Vijayakumar Mathaiyan
Jeju National University

M. Ramesh
Kumaraguru College of Technology

Dong Won Jung
Jeju National University

CONTENTS

9.1	Introduction about Nanocomposite	140
9.2	Literature Survey	141
	9.2.1 Summary	142
9.3	Methodology Used and Its Validation	142
	9.3.1 Experimental Testing	142
	9.3.1.1 Materials Used and Dimensions of Specimen	142
	9.3.1.2 Preparation of Test Specimen	143
	9.3.1.3 Testing Results	143
	9.3.2 Theoretical Calculation	146

           9.3.2.1 Investigation I 146
           9.3.2.2 Investigation II 146
     9.3.3 Finite Element Analysis of Nanocomposites 146
           9.3.3.1 Problem Formulation in FEA 146
           9.3.3.2 Conceptual Design 147
           9.3.3.3 Discretization 147
           9.3.3.4 Analysis Description and Its Control 147
           9.3.3.5 Boundary Conditions 149
           9.3.3.6 Governing Equations Used in FEA 150
           9.3.3.7 Grid Convergence Study 153
           9.3.3.8 Computational Structural Analysis 154
           9.3.3.9 Validational Investigations 154
9.4 Results and Discussion 156
     9.4.1 Details about Models Used 156
     9.4.2 Structural Simulations on SWCNT-Based CFRP 156
     9.4.3 Structural Simulations on MWCNT-Based CFRP 159
     9.4.4 Comparative Analysis of All the Deformations 162
     9.4.5 Comparative Analysis of All the Strain Energies 162
     9.4.6 Comparative Analysis of All the Stress Values 162
9.5 Conclusions 163
References 164

9.1 INTRODUCTION ABOUT NANOCOMPOSITE

Nanocomposites are the upgraded form in which the reinforcement, matrix, and filler all play a key role in the provision of properties and their enhancement. Nowadays, property enhancement is a primary investigating domain in composite materials, which is executed with the help of variations in the constituents of the composite. All the ingredients of the composites such as reinforcement, matrices, core materials, and fillers are undergoing investigation in order to increase the property of composites. In the case of reinforcement, the modifications for enhancement of the property vitally focus on the orientation and the types of reinforcement. When it comes to matrices, the primary focal point for enhancement is delamination, types of fillers, and their properties. Finally, with regard to core materials, the nature of the core materials plays a predominant role in the estimation of property [1]. Honeycombs and foams are perfect options to act as core materials for composites. Ceramics, nanomaterials, and teflons are the good subordinates of matrix for enhancing the characteristics of composites without affecting their lifetime. In this work, the tensile properties of the different nanocomposites are investigated by using advanced numerical simulation, which is based on finite element analysis (FEA) formulation. In this structural simulation, two different types of nanocomposites are used, in which carbon fiber and epoxy resin are fixed and served as reinforcement and additives, respectively. But one modification is executed at the mixture level, which means two different types of carbon nanotubes (CNTs) such as single-walled carbon nanotubes (SWCNTs) and multi-walled carbon nanotubes (MWCNTs) are implemented in the comparative structural analysis [3]. Carbon fiber–reinforced polymer (CFRP) composite is perfect to tackle aerospace applications, especially they are implemented in mechanical flight control system, empennage, and surface of aeroplane. In order to increase the output reliability of numerical simulation, experimental testing is done for validation purpose in this work [7].

9.2 LITERATURE SURVEY

Guptaa and Harsha [15] involved the CNTs as the primary agent in the polymer composites and thereby executed the FEA-based simulations in which the formulation in FEA, mechanical properties of the materials, types of meshes used, and elemental data implemented in the formation of composites were supported soundly in the current work in order to execute their FEA methodology. Mehar and Panda [17] analyzed the bending behavioral studies of nanocomposites, which were totally different from the nature of the present article; however, the primary theoretical principle information, the usage of computational tool, and mechanical characteristics of nanocomposites were guided in order to attain the preliminary stage. Especially the author of the present article initially struggled to analyze the nanocomposite behavior through ANSYS because of its reliability attainment. But the aforesaid reference provided a clear view on the usage of computational structural methodology and thereby in this work finalized to analyze the tensile behavior of nanocomposites with the help of ANSYS Workbench 16.2. Aubad et al. [19] investigated the structural analyses of hybrid laminate composite, in which MWCNTs, epoxy resin, and carbon fiber/Kevlar fiber were used as leading elements. Also, the FEA-based simulation and experimental testing were exploited to analyze the structural behavior of hybrid laminate composite. In the present comparative study, MWCNTs are involved as one of the primary materials; therefore, the needful data of MWCNTs and its analyzing procedures were extracted from this reference. Particularly, the mechanical properties of MWCNTs, the manufacturing process involved in the nanocomposite construction, and the advanced FEA simulation procedures for nanocomposite test specimen were clearly understood. Du et al. [20] reviewed the status of current need and the problems associated with CNT-based polymer composites, and they found that the nanocomposites have multidisciplinary advantages such as high thermal conductivity, better resisting force against tensile strength, good electrical conductivity, and high strength-to-weight ratio [2]. Because of these wide advantages, the nanocomposites have various industrial applications, for instance, thermal interface devices, optical instruments, electric equipment, and utilization materials for electromagnetic energy. The authors also studied the problems existing in the CNT-based composites: load transformation and its enhancement, and issues in nanocomposite construction in a right manner were found to be major complexities in CNT implementation. Finally, CNT's execution in polymer composites with respect to its amount of weight percentage, types of matrices used, types of manufacturing processes used, environmental types, etc. was analyzed [4]. The pros and cons have been investigated for all the perspectives. The above review articles contributed a huge amount of data to the present research work, especially the weight percentage level of filler to be added, manufacturing methodology for better production of nanocomposites, and suitable matrix implementation. Gojny et al. [8] worked on the comparative fracture analyses of nanocomposites through experimental testing, wherein all the CNTs have been employed. The present work compares the tensile strength in-between SWCNTs and MWCNTs, extracting major support from the above reference. Rubel et al. [22] reviewed the agglomeration effect of reinforced composites in the presence of CNTs, in which the nature and side effects on both reinforcement and matrices with the addition of CNTs were clearly explained [5].

9.2.1 SUMMARY

In most of the cases, the filler addition and its property enhancement with respect to its weight percentage inclusion in the composite-based investigations were completed with the support of experimental tests. The reasons behind the huge involvement of experimental test are its good reliability in the outcomes and the generation of test specimen in a controlled manner [6]. But in this work, the FEA-based computational simulations are used in the investigation of nanocomposites under tensile load conditions with a huge support from these aforementioned references. FEA-based nanocomposite analyses face two kinds of foremost difficulties, which make the outcomes unreliable. The two major difficulties principally involved in the FEA simulations are complexity in the construction of nanocomposite with its filler weight inclusion and obtaining boundary conditions needed for this complicated analyses. Thus, the strongest help is needed for these complicated analyses and hence the standard literature survey helped soundly in order to solve this structural simulation in an efficient manner [10]. Chiefly, the following conditions are obtained from the standard literatures: primary mechanical properties of MWCNTs and SWCNTs, good ranges of weight inclusion of filler in percentages, the manner in which CNTs are added in the perspective of computational analyses, external loading conditions, and support types. In this article, universal testing machine (UTM)-based experimental test is also engaged for the purpose of validation of computational results. In the experimental test phase, the following data are important: test specimen construction methodology, suitable environmental conditions, and lading details that are extracted from the standard literatures [13].

9.3 METHODOLOGY USED AND ITS VALIDATION

9.3.1 EXPERIMENTAL TESTING

9.3.1.1 Materials Used and Dimensions of Specimen

Experimentally, carbon fiber is short-listed for reinforcement, which plays a predominant role in the load-carrying function. Fundamentally, carbon fibers are dependent on graphene, which is primary element of carbon. Dry fabric and prepreg are the major available forms of carbon fiber. The natural color of carbon fiber is gray or black. Epoxy resin is used as a matrix, which is basically fit for all the available forms of carbon fiber. Epoxies come under thermosetting resin, and the viscosities of epoxy are available in all the stages from liquid to solid. Apart from carbon fiber and epoxy, CNTs contribute a lot in this work. A CNT is a tube-shaped material made of carbon, having a diameter that is measured on the nanometer scale [1]. In this experimental testing, MWCNT is used as mixtures in the CFRP for property enhancement. Five percent content of MWCNT is added to epoxy and then test specimen processes are

Comparative Numerical Analyses

FIGURE 9.1 Materials used for fabrication.

executed as per the standard procedure. The specifications for the design considerations of tensile test are followed as per the ASTM Standard (D3039), in which the length, width, and thickness of the specimen are 230, 25, and 5 mm, respectively. Figure 9.1 shows the preparation of nanocomposite for testing.

9.3.1.2 Preparation of Test Specimen

The contents used in this nanocomposite are 60% of carbon fiber (263 g) as reinforcement, 40% of epoxy resin with hardener, and 5% of MWCNTs as a matrix. After successfully finalizing the content, the fabrication process is finalized. Comparatively, compression molding process is more suitable for the nanocomposite generation because of its output reliability and user-friendly nature, which make it fit for the construction of all kinds of composite material. The common and general procedures are followed in this nanocomposite construction, in which 3 psi pressure is used for the compression purpose [1]. Figure 9.2 shows the typical process involved in the compression molding, and Figure 9.3 reveals the output of compression molding process. And then the final products are shown in Figure 9.4.

9.3.1.3 Testing Results

Mechanical test is generally used to provide a complicated experience to test specimen, which provides the tackling technique in order to overcome complicated environments. Tensile, bending, and impact tests are primary evaluation methodologies involved in the structural analysis, in which tensile test is the base and universal engineering evaluation methodology to attain and analyze the structural parameters. The important structural parameters are modulus of elasticity, % area of reduction, yield strength, % elongation, and ultimate strength. The working environment of the tensile test is loaded in axial direction at one end and the other end of the test specimen is fixed at UTM jars. The detailed pictorial representations are shown in Figure 9.5. The known values of the tensile test are gauge length and perpendicular area of the test specimen, which supported a lot in the calculation of stress and strain of test material [11]. The tensile tests are conducted in the room temperature and the results are noted, which is also shown in Figure 9.6, and the comprehensive data of this test outputs are listed in Table 9.1.

FIGURE 9.2 Compression molding process.

FIGURE 9.3 Test specimen preparation.

FIGURE 9.4 Test specimens.

Comparative Numerical Analyses 145

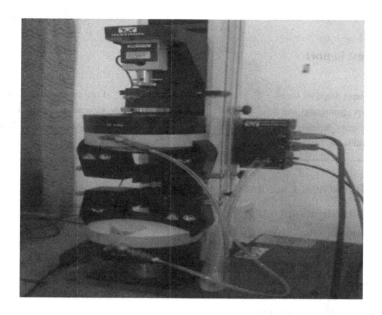

FIGURE 9.5 Universal testing machine with test specimen.

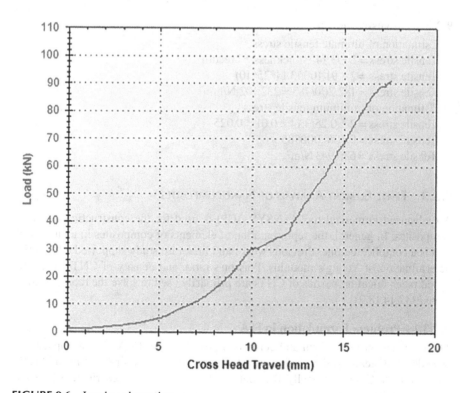

FIGURE 9.6 Load vs elongation.

TABLE 9.1
Tensile Test Report

Input Data	Output Data
Specimen shape: dog bone shape	Load at yield: 91.10 kN
Specimen type: nanocomposite	Elongation at yield: 1.5300 mm
Specimen description: carbon fiber+ MWCNTs	Yield stress: 581.600 N/mm^2
Specimen width, thickness, length: 25, 10, 175 mm	

9.3.2 THEORETICAL CALCULATION

9.3.2.1 Investigation I
Estimation of ultimate stress
Stress = load/cross-sectional area
Normal stress = 91,100/(25 × 10) = 91,100/250
Normal stress = 364.4 N/mm^2
Estimation of minimum stress
Stress = load/cross-sectional area
Normal stress = 0.25/0.00025
Normal stress = 1,000 N/m^2

9.3.2.2 Investigation II
Estimation of ultimate tensile stress
Tensile stress = 2 F/3.14 * thickness * breadth
Tensile stress = 2 * 91,100/(3.14*25*10)
Tensile stress = 182,200/785 = 232.102 N/mm^2
Estimation of minimum tensile stress
Tensile stress = 2 * 0.25/3.14 * 0.01 * 0.025
Tensile stress = 0.50/0.000785
Tensile stress = 636.943 N/m^2

9.3.3 FINITE ELEMENT ANALYSIS OF NANOCOMPOSITES

An advanced numerical tool (ANSYS ACP) is used for the construction of nanocomposites. In general, the representation of elements of composites in a numerical tool is a complicated one. Previous works and literature study supported a lot in the determination of Young's modulus, Poisson's ratio, and density of CNTs. The collected mechanical properties of CNTs are primarily used to solve the representation issue [9,12,14,18,21].

9.3.3.1 Problem Formulation in FEA
The involvement of uniform and controlled procedures in FEA has the capacity to provide most steadfast end results, which is a mandatory requirement in the FEA's research field. Computationally, the controlled procedures are described as follows: conceptual design of a test specimen, finite element model of a test specimen,

Comparative Numerical Analyses

description of structural analysis used, boundary conditions are implemented, control involved in the structural analysis, and employment of governing equations in FEA and grid convergence study. Complicated works must follow these controlled procedures in order to attain good results even though computer-based simulations are implemented. Thus, the comparative numerical analyses of nanocomposites under tensile load are completed with the help of aforementioned stages.

9.3.3.2 Conceptual Design

Computational model is the basic platform of numerical simulation, which represents the real-time object. In general, conceptual design is a process that involves the determination and construction of three-dimensional data of a test specimen. The perfect resemblance of a test specimen is predominant and a preliminary requirement in a numerical simulation. Therefore, the computational model and its preliminary steps are unavoidable and represent the major stage in computational structural simulation. This test model execution is attained through two different modes in the first mode, the model is completed by import facility with the support of igs/stp format design files, and in the second mode, the model is completed by the facility of design modeler of the intended FEA solver. Basically, complicated shapes such as aircraft wing, hydro-rotors, and wind turbine have been imported by the aforesaid importing facility and the simple geometries are executed by the inbuilt design tool. In this article, dog bone–shaped test specimen is vitally used as a computational model because ASTM D-3039 supported a lot in this fine construction by using ANSYS Workbench 16.2 [9,14]. The final model of test specimen is revealed in Figure 9.7.

9.3.3.3 Discretization

Basically, discretization is the conversion process of a complete test model into segregated finite element models. The segregation happens in two ways: structured and unstructured. Mostly, structured grid formations are preferable to implement in the discretization because of its highly reliable output with low grid constructional time. Also, the requirements of discretized parts such as nodes and elements are in the medium level only. Whereas in an unstructured case, the time consumed f or the quality grid constructions is higher, which increases the computational time and reliable output production. Therefore, in this work, structured meshes are generated from the test model (dog bone–shaped specimen) with the support of ANSYS Mesh Tool 16.2 and the discretized model is shown in Figure 9.8 as sample case. Generally, the ANSYS-based FEA tool has the facility to construct quality structured mesh on the standard test models, but in the complicated test models such as turbine and wing, additional facilities are required in order to generate a fine mesh. In this work, the mesh quality obtained is 0.9, which provides good outcomes. Despite attaining so much of good quality, this research article also analyzed the grid convergence study for the purpose of complete reduction of mesh errors.

9.3.3.4 Analysis Description and Its Control

The nature of this research work is fundamental analyses of tensile property of various nanocomposites, so the steady and static behavior–based solver is used. In this fundamental investigation, the following structural parameters are considered and

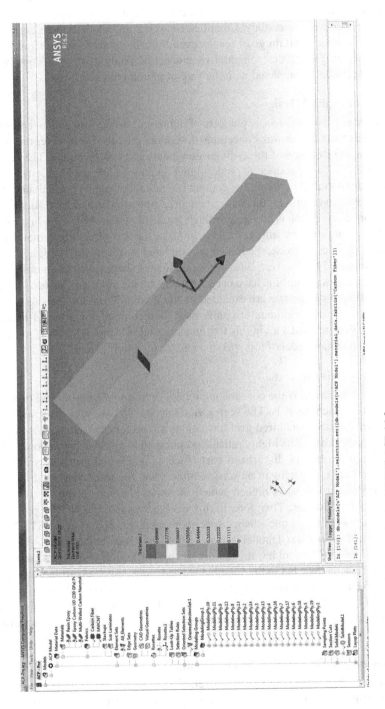

FIGURE 9.7 Pure composite generation in ANSYS ACP 16.2.

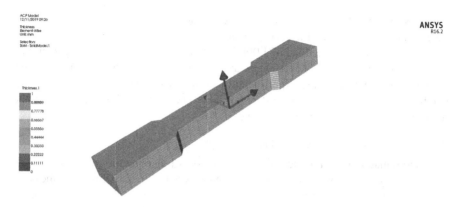

FIGURE 9.8 Discretized structure of test specimen.

they are used for the selection process: (1) deformations generated on the various test models, (2) equivalent stresses induced inside the dog bone–shaped model, and (3) strain energies developed in the internal structure of the nanocomposite–based computational models. Thus, the estimation of displacement, strain, and stress is important under tensile load, where a simplified version of FEA-based governing equations is capable of predicting these outputs. The 15 important equations are listed out as equations 9.1–9.16. The first three equations represent the tensile load implementation in all the three directions. And next 12 equations are connectivity equations in-between displacement, strain, and stress. In general, the Newton's second law is the fundamental platform for this external load–based equilibrium equation in all directions, where acceleration is assumed to be zero because the nature of this work is static. Also, the stiffness-based approach is followed in this work due to the clear availability of magnitude of external loading conditions.

9.3.3.5 Boundary Conditions

Computationally, the boundary conditions are the key factor to initiate the numerical simulations in which the stress induction and deformations are created based on the nature of the boundary conditions. Generally, initial conditions and boundary values are available as subgroups in the boundary conditions; moreover, the mechanical properties such as Poisson's ratio, density, Young's modulus, and thermal conductivity and three-dimensional geometrical properties come under initial input conditions. When it comes to boundary values, the forces acting on the test models, supports in the test models, cross-sectional areas, connectivity, etc. play crucial roles. In this work, three important boundary conditions drastically affected the outcomes, which are mechanical properties of various nanocomposites, application of external tensile load and its direction, and finally the type of support provided. Standard literature survey supported soundly in attaining the different properties of nanocomposites, especially strain ratio and modulus are predominantly collected and used in these comparative analyses. Considering external load phase, two kinds of tensile loads are implemented in two different environments. The first environment is the validational study of computational structural analysis, where the experimental output of

72,700 N is applied as ultimate tensile load at one end. After the validation comes the second environment, which is comparative analyses in-between nanocomposites, where the common tensile load of 1,000 Pa is provided at one of the cross-sectional areas. Finally, the fixed support is gen at the other cross-sectional area of the test model, which represents the experimental loading conditions. The entire boundary condition applied is revealed in Figure 9.9, where pressure load acts at one end and fixed support is attached at the other end.

9.3.3.6 Governing Equations Used in FEA

In general, mathematical modeling consists of governing equations defined in a field and boundary conditions provided at the boundaries of the area. In this work, the composites are primary platforms, so two more important equations need to be included in order to provide the required and acceptable output. The important equations are 3-D Hooke's law equation and strain-displacement relationships. Finally, 15 subequations are predominantly used in these FEA-based frictional stress calculations.

Force balance in X-direction

$$\frac{\partial \sigma_x}{\partial x} + \frac{\partial \tau_{xy}}{\partial y} + \frac{\partial \tau_{xz}}{\partial z} + F_x = 0 \tag{9.1}$$

Force balance in Y-direction

$$\frac{\partial \tau_{xy}}{\partial x} + \frac{\partial \sigma_y}{\partial y} + \frac{\partial \tau_{yz}}{\partial z} + F_y = 0 \tag{9.2}$$

Force balance in Z-direction

$$\frac{\partial \tau_{xz}}{\partial x} + \frac{\partial \tau_{yz}}{\partial y} + \frac{\partial \sigma_z}{\partial z} + F_z = 0 \tag{9.3}$$

Normal strain in X-direction

$$\varepsilon_x = \frac{\partial u}{\partial x} \tag{9.4}$$

Normal strain in Y-direction

$$\varepsilon_y = \frac{\partial v}{\partial y} \tag{9.5}$$

Normal strain in Z-direction

$$\varepsilon_z = \frac{\partial w}{\partial z} \tag{9.6}$$

Comparative Numerical Analyses

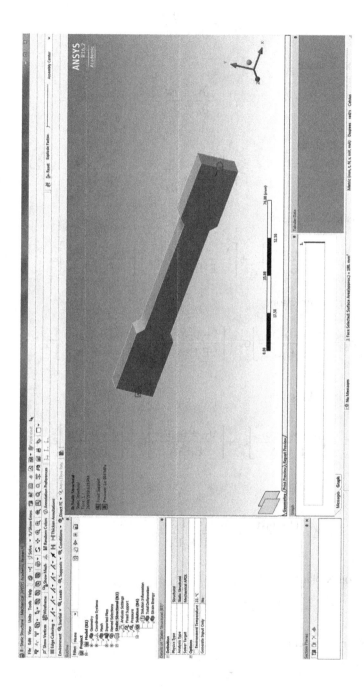

FIGURE 9.9 Boundary conditions applied.

Shear strain on X-plane

$$\gamma_{xy} = \frac{\partial u}{\partial y} + \frac{\partial v}{\partial x} \tag{9.7}$$

Shear strain on Y-plane

$$\gamma_{yz} = \frac{\partial v}{\partial z} + \frac{\partial w}{\partial y} \tag{9.8}$$

Shear strain on Z-plane

$$\gamma_{zx} = \frac{\partial w}{\partial x} + \frac{\partial u}{\partial z} \tag{9.9}$$

Normal stress in X-direction

$$\sigma_x = \left[\frac{1-\upsilon_{23}\upsilon_{32}}{E_2 E_3 \Delta}\right]\varepsilon_x + \left[\frac{\upsilon_{21}+\upsilon_{23}\upsilon_{31}}{E_2 E_3 \Delta}\right]\varepsilon_y + \left[\frac{\upsilon_{31}+\upsilon_{21}\upsilon_{32}}{E_2 E_3 \Delta}\right]\varepsilon_z \tag{9.10}$$

Normal stress in Y-direction

$$\sigma_y = \left[\frac{\upsilon_{21}+\upsilon_{23}\upsilon_{31}}{E_2 E_3 \Delta}\right]\sigma_x + \left[\frac{1-\upsilon_{13}\upsilon_{31}}{E_1 E_3 \Delta}\right]\sigma_y + \left[\frac{\upsilon_{32}+\upsilon_{12}\upsilon_{31}}{E_1 E_3 \Delta}\right]\varepsilon_z \tag{9.11}$$

Normal stress in Z-direction

$$\sigma_z = \left[\frac{\upsilon_{31}+\upsilon_{21}\upsilon_{32}}{E_2 E_3 \Delta}\right]\varepsilon_x + \left[\frac{\upsilon_{32}+\upsilon_{12}\upsilon_{31}}{E_1 E_3 \Delta}\right]\varepsilon_y + \left[\frac{1-\upsilon_{12}\upsilon_{21}}{E_1 E_2 \Delta}\right]\varepsilon_z \tag{9.12}$$

Where

$$\Delta = \frac{(1-\upsilon_{12}\upsilon_{21}-\upsilon_{23}\upsilon_{32}-\upsilon_{13}\upsilon_{31}-2\upsilon_{21}\upsilon_{32}\upsilon_{13})}{E_1 E_2 E_3}$$

Shear stress on X-plane

$$\tau_{xy} = [G_{12}]\gamma_{xy} \tag{9.13}$$

Shear stress in Y-direction

$$\tau_{yz} = [G_{23}]\gamma_{yz} \tag{9.14}$$

Shear stress in Z-direction

9.3.3.7 Grid Convergence Study

In recent times, the grid construction process is made easier because of the huge development in the advanced computational tools. Also, the unavoidable facilities in the discretization are moved as optional service; thus, the formation of unstructured mesh is increasing in an abnormal manner. With this easiest mesh, the researcher will be able to analyze the computational problems, which increased the untrustworthiness in the results. Therefore, extra effort is required in the discretization in which gird convergence study and numerical sensitivity assessment based on external loads are the topmost implemented techniques. In this work, the grid convergence study is used for the reduction of discretization-dependent error, wherein the grid convergence study is a conventional method in computational investigations. Fundamentally, grid convergence study is the optimization process in which the optimistic constituents are nodes and elements of mesh cases. Based on the constructional qualities, four various grid cases are used from coarse mesh to fine with face set-up mesh. The low elemental quality–based grids that are generated come under coarse category and the high elemental quality–based grids come under fine with face mesh set-up.

The discretization-based errors are reduced through the use of conventional methods, that is, grid independence method. In this regard, four different mesh cases are formed and its complete details are listed in Table 9.2. The discretized structures are revealed in Figures 9.10 and 9.11, where the coarse mesh is shown in Figure 9.10 and fine mesh is shown in Figure 9.11. At last, the aforesaid boundary conditions are used

TABLE 9.2
Various Mesh Cases with Nodes and Elements

Type	Cases	Nodes	Elements
Coarse	Case 1	208	1,025
Medium	Case 2	527	2,698
Fine	Case 3	1,254	10,257
Fine with face set-up	Case 4	2,587	23,654

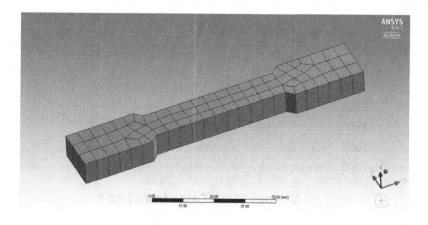

FIGURE 9.10 A wireframe model view of coarse mesh.

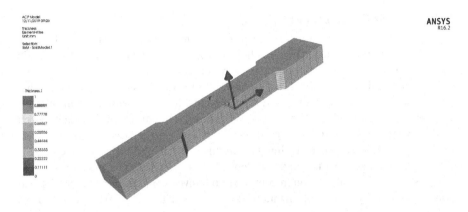

FIGURE 9.11 A wireframe model of fine mesh.

in all the four cases for the SWCNT-based nanocomposites and the comprehensive graph is shown in Figure 9.12. From Figure 9.12, it is clearly understood that case 3 (fine mesh) is better than case 4 (fine with face set-up).

9.3.3.8 Computational Structural Analysis

The validational-based structural analyses are executed with the load of 91,100 N at one end and the other end is provided with fixed support. The maximum tensile load is estimated from experimental test. Figures 9.13 and 9.14 reveal the strain energy and total deformation variations in the test specimens, respectively, in which the SWNCT-carbon fiber–based composite simulation is carried out only for validation [9,12,14,18].

9.3.3.9 Validational Investigations

In this research article, two kinds of validation are executed because of the working complexity of this intended comparative analysis. Generally, filler has tremendous impact on property enhancement in the composites; thus, the filler-based comparative analyses play a vital role in nanocomposites. First validation is based on an

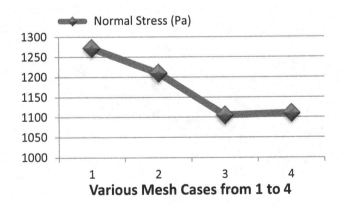

FIGURE 9.12 Comparative FEA results with mesh cases.

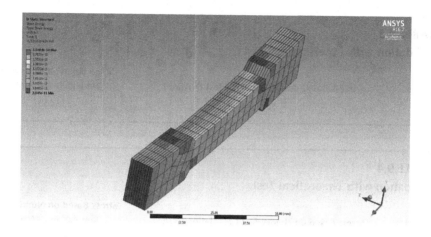

FIGURE 9.13 Strain energy variations.

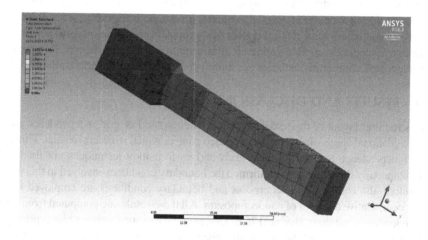

FIGURE 9.14 Deformed structure.

experimental test with computational structural simulations. Firstly, the UTM-based experimental tests are conducted at room temperature, in which the maximum tensile stress, deformation, and maximum tensile load are noted. Secondly, the computational structural analyses are executed for CFRP prepreg with MWCNTs under experimentally obtained load and so the structural simulations are computed. The entire comprehensive results are compared and listed in Table 9.3. From Table 9.3, it is understood that the end results are closest with each other; therefore, the implemented grids and boundary conditions are validated, which provided more confidence to implement these results in real-time applications. Apart from the experimental validation, the standard theoretical-based validation is also executed in this work, in which the common loading condition is provided as 0.25 N, which is considered as the second validation. The comparative investigations of theoretical

TABLE 9.3
Validation with Experimental Test

Sl. No	Stress Based on Experimental Test (N/mm^2)	Stress Based on Numerical Simulation (N/mm^2)
1	581.600	538.653

TABLE 9.4
Validation with Theoretical Test

Sl. No	Stress Based on Theoretical Calculation (N/m^2)	Stress Based on Numerical Simulation (N/m^2)
1	1,000	1,105.08

and computational structural analyses are computed and the results are listed in Table 9.4. The error percentage is also obtained with the help of data from Table 9.4, and the value is less than 10%; therefore, the proposed computational studies and its primary procedures are also validated at minimum loading conditions. Henceforth, the suggested comparative studies are more perfect for real-time applications.

9.4 RESULTS AND DISCUSSION

The structural layout of this article is primarily composed of grid independence study and validational approaches for these proposed computational structural results. Through grid independence test, the suitable grids and its formation techniques are finalized. Simultaneously the validations confirmed the boundary conditions involved in this work. Therefore, the confirmed mesh process and boundary conditions are employed in all these comparative analyses of nanocomposites. All these results are computed from minimum loading conditions, that is, 0.25 N. Deformation and strain energy play a principal role in the selection of suitable nanocomposite materials of this comparative analysis.

9.4.1 Details about Models Used

Totally ten different computational models are finalized and implemented for this research work. The complete data of all the computational models are listed in Table 9.5, in which the composite laminate percentages vary from 90% to 99% and the different CNTs vary from 1% to 10%. All these model constructions were finalized from standard literature survey [16].

9.4.2 Structural Simulations on SWCNT-Based CFRP

In the first case, SWCNT-based nanocomposites are analyzed. The sample structural variation is revealed in Figures 9.15 and 9.16, in which the deformed structure under tensile load is shown in Figure 9.15 and the variations in strain energy is shown in Figure 9.16.

TABLE 9.5
Information about Various Test Models Involved in This Comparative Analysis

	Composite Based on SWCNTs		Composite Based on MWCNTs	
Model No	Laminate Contents	SWCNT Contents	MWCNT Contents	Laminate Contents
1	9.9	0.1	0.1	9.9
2	9.8	0.2	0.2	9.8
3	9.7	0.3	0.3	9.7
4	9.6	0.4	0.4	9.6
5	9.5	0.5	0.5	9.5
6	9.4	0.6	0.6	9.4
7	9.3	0.7	0.7	9.3
8	9.2	0.8	0.8	9.2
9	9.1	0.9	0.9	9.1
10	9.0	1.0	1.0	9.0

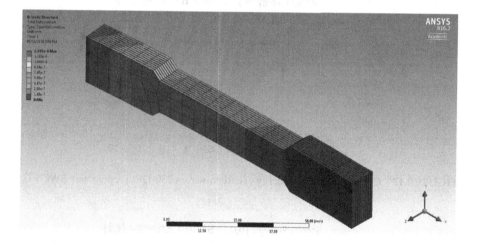

FIGURE 9.15 Deformed shape of a test specimen.

The comparative strain energy analyses in-between ten SWCNT-based nanocomposite models are comprehensively shown in Figures 9.17 and 9.18, in which model 10 is the perfect combination. In Figure 9.17, the carbon fiber is used as carbon-woven 230-GPa prepreg, and in Figure 9.18, the carbon fiber is used as carbon-woven 230-GPa wet.

The comparative deformation analyses in-between ten SWCNT-based nanocomposite models are comprehensively shown in Figures 9.19 and 9.20, in which model 10 and model 1 are perfect combinations. In Figure 9.19, the carbon fiber is used as carbon-woven 230-GPa prepreg, and in Figure 9.20, the carbon fiber is used as carbon-woven 230-GPa wet.

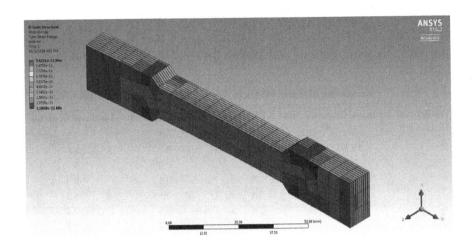

FIGURE 9.16 Distributions of strain energy.

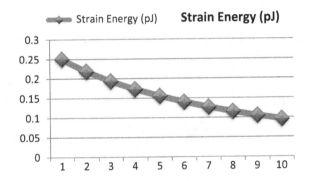

FIGURE 9.17 Comparative strain energy of carbon-woven 230-GPa prepreg with SWCNTs.

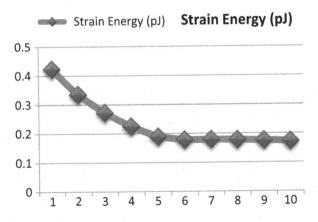

FIGURE 9.18 Comparative strain energy of carbon-woven 230-GPa wet with SWCNTs.

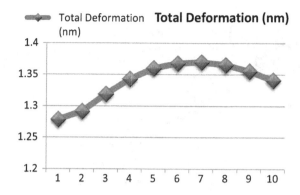

FIGURE 9.19 Comparative deformation of carbon-woven 230-GPa prepreg with SWCNTs.

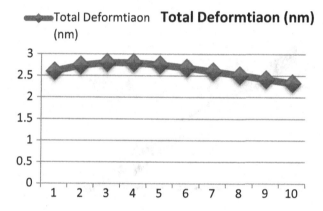

FIGURE 9.20 Comparative deformation of carbon-woven 230-GPa wet with SWCNTs.

9.4.3 Structural Simulations on MWCNT-Based CFRP

The structural simulations of MWCNT-based CFRP are executed, in which the cantilever structure–resembled boundary conditions are followed. The important evaluation parameters such as strain energy and deformation are completed, which are shown in Figures 9.21 and 9.22.

The comparative strain energy analyses of ten MWCNT-based nanocomposite models are comprehensively shown in Figures 9.23 and 9.24, in which model 10 is the perfect combination. In Figure 9.23, the carbon fiber is used as carbon-woven 230-GPa prepreg, and in Figure 9.24, the carbon fiber is used as carbon-woven 230-GPa wet.

The comparative deformation analyses of ten MWCNT-based nanocomposite models are comprehensively shown in Figures 9.25 and 9.26, in which model 10 and model 7 are the perfect combinations. In Figure 9.25, the carbon fiber is used as carbon-woven 230-GPa prepreg, and in Figure 9.26, the carbon fiber is used as carbon-woven 230-GPa wet.

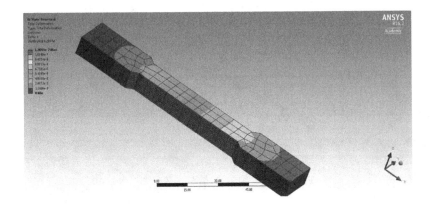

FIGURE 9.21 Variations in deformation of test specimen.

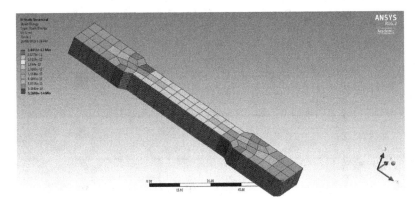

FIGURE 9.22 Variations in strain energy.

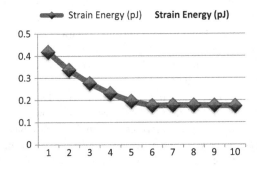

FIGURE 9.23 Comparative strain energy analysis of carbon-woven 230-GPa prepreg with MWCNTs.

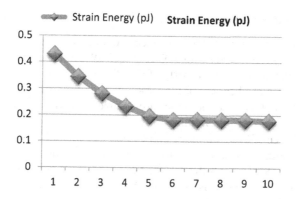

FIGURE 9.24 Comparative strain energy analysis of carbon-woven 230-GPa wet with MWCNTs.

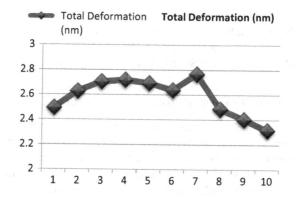

FIGURE 9.25 Comparative deformation analysis of carbon-woven 230-GPa prepreg with MWCNTs.

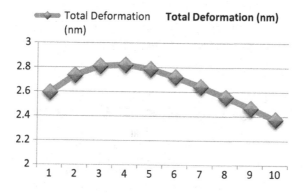

FIGURE 9.26 Comparative deformation analysis of carbon-woven 230-GPa wet with MWCNTs.

9.4.4 Comparative Analysis of All the Deformations

The comparative structural analyses are executed for both the cases, in which carbon fiber–based epoxy resin added with MWCNT and SWCNT primarily plays a fantastic role in the enhancement of property of CFRP. Also, in the comparative analysis, two different types of fibers are used: carbon-woven 230-GPa prepreg and carbon-woven 230-GPa wet. All the four cases are completed, and the comparative results are noted in Figures 9.27–9.29. Tables 9.1 and 9.2 provide the results of SWCNTs with carbon-woven 230-GPa prepreg and carbon-woven 230-GPa wet, respectively. In all the four cases, the evaluation parameters considered are strain energy and displacement. Figure 9.27 reveals the deformations of all the four cases, in which CFRP prepreg with SWNCT models is fit to resist high amount of loads because they react low when compared to others.

9.4.5 Comparative Analysis of All the Strain Energies

Figure 9.28 shows the strain energies of all the four cases, in which CFRP prepreg with SWNCT models is fit to resist high amount of loads because they react low when compared to others. Thus, from the first phase of this work, it is clearly shown that CFRP prepreg with SWNCT models are good to handle tensile loads.

9.4.6 Comparative Analysis of All the Stress Values

From Figures 9.27 and 9.28, CFRP prepreg with SWNCT models is better; thus, the computational structural analysis is extended for stress estimation. The comparative stress results are shown in Figure 9.29, in which the primary materials are CFRP prepreg with SWCNTs and CFRP wet with SWCNTs. Finally, SWCNTs equipped with carbon-woven 230-GPa prepreg is more perfect for tensile applications because of its low strain energy induction, low stress generation, and low deformed structure.

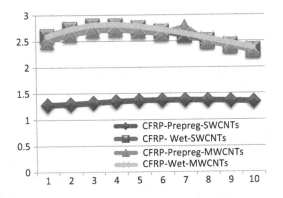

FIGURE 9.27 Comparative analysis of all the deformations.

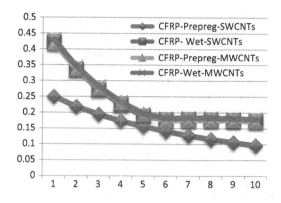

FIGURE 9.28 Comparative analysis of all the strain energies.

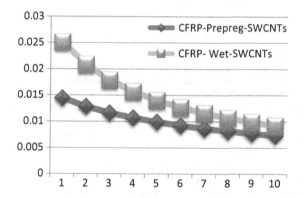

FIGURE 9.29 Comparative analysis of all the stress values.

9.5 CONCLUSIONS

All the fundamental data such as three-dimensional information, mechanical properties of used materials, boundary conditions, and selection of main and sub elements of nanocomposites of this work are obtained from the standard literature survey. The preprocessing of this simulation methodology is completed with the help of ANSYS Design Modeler 16.2, ANSYS Mesh Tool 16.2, and ANSYS ACP 16.2. The solving process of this work is executed based on stiffness approach in ANSYS Static Structural tool 16.2. The fundamental need of this work for evaluation purpose is total deformation and strain energy that provided the path for stiffness-based selection and solution. Finally, the comparative analyses are executed for various nanocomposites with various mixtures, and thereby the validation part is executed for numerical results of 5% content of MWCNT-based nanocomposites with the help of standard experimental testing. The validation shows that the advanced one-way coupled methodology used in this work is more fit and reliable to solve composite problems. At last, the SWCNT structural results are better than MWCNT structural results.

REFERENCES

1. G. Raj Kumar et al., The mechanical characterization of carbon fiber reinforced epoxy with carbon nanotubes, *International Journal of Mechanical and Production Engineering Research and Development*, 9, 1, 243–255 (2019).
2. H. W. Zhou, L. Mishnaevsky Jr., H. Y. Yi, Y. Q. Liu, X. Hu, A. Warrier, G. M. Dai, Carbon fiber/carbon nanotube reinforced hierarchical composites: Effect of CNT distribution on shearing strength, *Composites Part B*, 88, 201–211 (2016), http://dx.doi.org/10.1016/j.compositesb.2015.10.035
3. R. Udhaya Prakash, G. Raj Kumar, R. Vijayanandh, M. Senthil Kumar, T. Ram Ganesh, Structural analysis of aircraft fuselage splice. *IOP Conference Series: Materials Science and Engineering Journal*, 149, 1, 012127 (2017).
4. P. Zhao, G. Shi, Study of Poisson ratios of single-walled carbon nanotubes based on an improved molecular structural mechanics model, *CMC*, 22, 2, 147–168 (2011).
5. G. Raj Kumar, R. Vijayanandh, M. Senthil Kumar, S. Sathish Kumar, Experimental testing and numerical simulation on natural composite for aerospace applications, ICC 2017, *AIP Conference Proceedings*, 1953, https://doi.org/10.1063/1.5032892.
6. S. I. Yengejeh, S. Alieh Kazemi, A. Öchsner, Carbon nanotubes as reinforcement in composites: A review of the analytical, numerical and experimental approaches, *Computational Materials Science*, 136, 85–101 (2017), http://dx.doi.org/10.1016/j.commatsci.2017.04.023
7. G. Raj Kumar et al., Conceptual design and structural analysis of integrated composite Micro Aerial Vehicle, *Journal of Advanced Research in Dynamical and Control Systems*, 9, 14, 857–881 (2017).
8. F. H. Gojny, M. H. G. Wichmann, B. Fiedler, K. Schulte, Influence of different carbon nanotubes on the mechanical properties of epoxy matrix composites—A comparative study, *Composites Science and Technology*, 65, 2300–2313 (2005), doi:10.1016/j.compscitech.2005.04.021.
9. R. Vijayanandh, K. Naveen Kumar, M. Senthil Kumar, G. Raj Kumar, R. Naveen Kumar, L. Ahilla Bharathy, Material optimization of high speed micro aerial vehicle using FSI simulation, *Procedia Computer Science*, 133, 2–9 (2018).
10. M. Kulkarni, D. Carnahan, K. Kulkarni, D. Qian, J. L. Abot, Elastic response of a carbon nanotube fiber reinforced polymeric composite: A numerical and experimental study, *Composites: Part B*, 41, 414–421 (2010), doi:10.1016/j.compositesb.2009.09.003
11. M. Rajagurunathan, G. Raj Kumar, R. Vijayanandh, V. Vishnu, C. Rakesh Kumar, K. Mohamed Bak, The design optimization of the circular piezoelectric bimorph actuators using FEA, *International Journal of Mechanical and Production Engineering Research and Development*, 8, 7, 410–422 (2018).
12. R. Vijayanandh et al., Numerical study on structural health monitoring for unmanned aerial vehicle, *Journal of Advanced Research in Dynamical and Control Systems*, 9, 6, 1937–1958 (2017).
13. S. Yellampalli, *Carbon Nanotubes – Polymer Nanocomposites*, In Tech, Rijeka, Croatia.
14. R. Vijayanandh et al., Vibrational fatigue analysis of NACA 63215 small horizontal axis wind turbine blade, *Materials Today Proceedings*, 5, 2, 6665–6674 (2018), https://doi.org/10.1016/j.matpr.2017.11.323.
15. A. K. Guptaa, S.P. Harsha, Analysis of mechanical properties of carbon nanotube reinforced polymer composites using continuum mechanics approach, *Procedia Materials Science*, 6, 18–25 (2014).
16. K. Venkatesan et al., Comparative structural analysis of advanced multi-layer composite materials, *Materials Today: Proceedings*, (2019), https://doi.org/10.1016/j.matpr.2019.11.247.

17. K. Mehar, S. K. Panda, Elastic bending and stress analysis of carbon nanotube-reinforced composite plate: Experimental, numerical, and simulation, *Advances in Polymer Technology*, 37, 1643–1657 (2018), doi:10.1002/adv.21821.
18. R. Vijayanandh, K. Venkatesan, M. Ramesh, G. Raj Kumar, M. Senthil Kumar, Optimization of orientation of carbon fiber reinforced polymer based on structural analysis, *International Journal of Scientific& Technology Research*, 8, 11, 3020–3029, (2019).
19. M. J. Aubad, B. A. Abass, S. N. Shareef, Investigation of the effect of multi wall carbon nano tubes on the dynamic characteristics of woven kevlar/carbon fibers-polyester composites, *Materials Research* Express, 7, 015054 (2020), doi:10.1088/2053-1591/ab5d67
20. J.-H. Du, J. Bai, H.-M. Cheng, The present status and key problems of carbon nanotube based polymer composites, *eXPRESS Polymer Letters*, 1, 5, 253–273, (2007), doi:10.3144/expresspolymlett.2007.39.
21. M. Chwał, A. Muc, Design of reinforcement in nano and microcomposites, *Materials*, 12, 1474, 1–23 (2019), doi:10.3390/ma12091474.
22. R. I. Rubel, H. Ali, A. Jafor, M. Alam, Carbon nanotubes agglomeration in reinforced composites: A review, *AIMS Materials Science*, 6, 5, 756–780 (2019), doi: 10.3934/matersci.2019.5.756.

Index

aging 10
atomic energy microscope 20
atomic force microscopy 22, 43
autogenic pressure reaction 89, 90, 104
automotive industry 25, 35, 57, 58

cancer cells 10
carbon fiber 122, 140, 141, 142, 157, 157, 159
cast iron manifold 112, 114, 118, 119
categorization 3
characterization 19
characterization techniques 42, 63, 74
chromium 112, 114, 119
composite gear 135, 136
composites 136, 140, 141, 146
construction 3, 17, 32, 33, 35, 45, 53, 141, 156

dielectric 35, 36, 91, 96, 98, 105, 122, 124
differential scanning calorimetry (DSC) 43, 76, 78, 80
diffraction 62, 63, 64, 65, 66, 67, 68, 69, 71, 80, 91
dynamic mechanical thermal analysis (DMTA) 78, 79, 80

electron microscopy 44, 58, 68, 69, 70, 91

formulation 140, 141, 146
friction 54, 55, 121, 124, 126, 129, 131, 136, 150
FTIR 70, 80
FTIR spectroscopy 70

graphene 22, 23, 32, 33, 51, 55, 88, 89, 90, 94, 142

lightweight 22, 26, 31, 55
load 103, 104, 112, 123, 128, 129, 131, 142, 149

mechanical properties 17, 19, 33, 34, 40, 41, 42, 44, 45, 52, 56, 103, 163
microscopy 20, 22, 69, 70, 133, 136
MWCNTs 142, 142, 143, 155, 22, 56

nano coatings 51, 56
nano particles 46

neutron diffraction 65, 66, 67, 68
nickel 55, 57, 112, 114, 119
nuclear magnetic resonance 74

optimization 33, 40, 153

polyamide 6 (PA6) 122, 123, 124, 125, 126, 129, 136
polymer matrix 22, 25, 32, 36, 39, 40, 42, 43, 45, 47, 56, 57, 62, 69, 70, 77
polymers 6, 8, 11, 22, 33, 36, 44, 45, 62, 63, 64, 65, 68, 76, 77, 78, 80
pressure sensing 53

Raman spectroscopy 72, 73
reinforcement ratios 40, 42

scanning tunneling microscope 20, 21, 112, 113, 114, 118
SEM 112, 113, 114, 118
spectroscopy 70, 71, 72, 73, 74, 75, 76, 91, 94, 95
speed 9, 35, 40, 68, 91, 124, 128, 129, 131
strength 7, 20, 22, 23, 25, 26, 32, 33, 34, 36, 38, 40, 41, 42, 44, 45
SWCNTs 22, 140, 142, 162

Tafel 113, 114, 115
tensile test 143
TGA 42, 78, 80, 91, 92, 93, 105
thermal conductivity 52, 88, 89, 92, 98, 99, 105
thermogravimetric analysis 62, 78, 80, 91, 92, 100, 101
TMA 76, 79, 99, 105
transmission electron microscopy 68, 69

UV-visible absorption spectroscopy 71

water absorption 41
wear 124, 126, 127, 128, 129, 131, 133, 136

X-ray diffraction (XRD) 63, 64, 65, 80, 91, 94, 104

zirconium oxide (ZrO_2) 121, 133

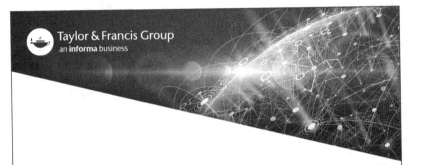

Printed in the United States
by Baker & Taylor Publisher Services